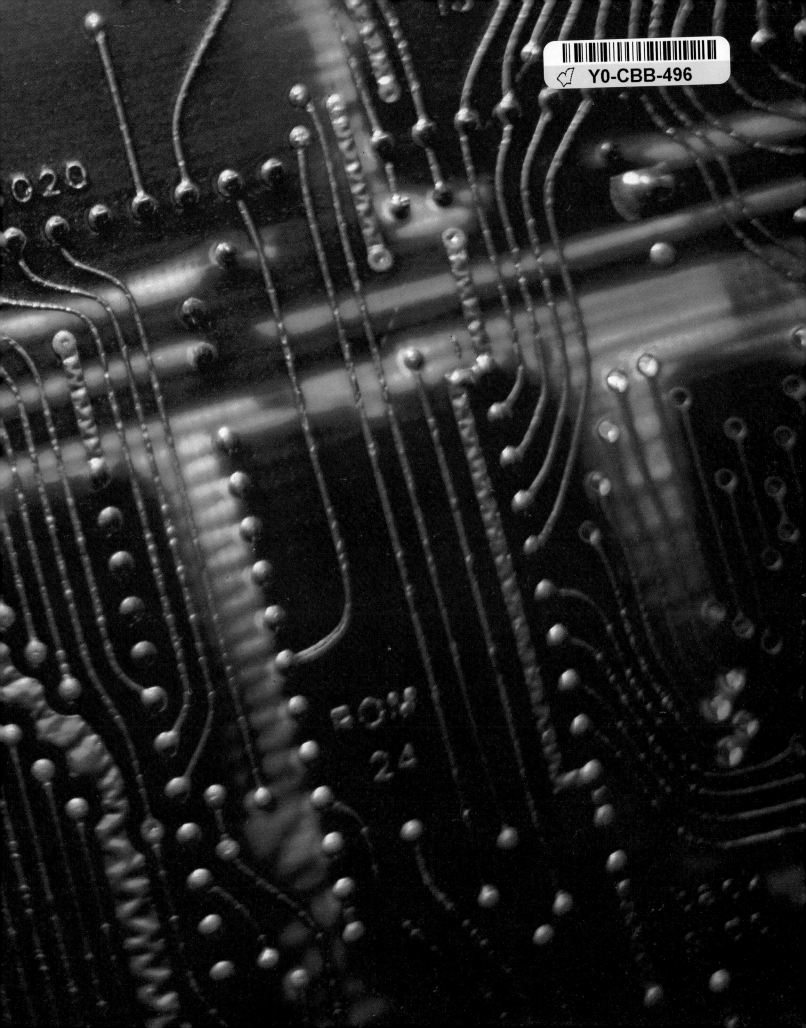

INSIDE MODERN TECHNOLOGY
HOW IT WORKS

GREENWICH HOUSE
New York

Editor: Quentin Deane
Designer: Brenda Morrison
Commissioning editor: Fran Jones
Production: Dennis Hovell

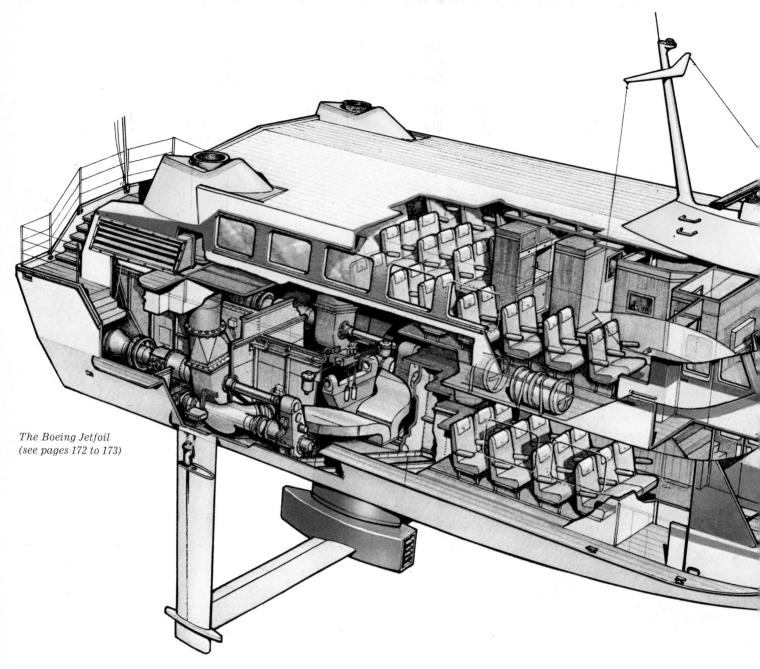

The Boeing Jetfoil
(see pages 172 to 173)

First English edition published by
Marshall Cavendish Limited.
This edition is published by Greenwich House.
Distributed by Crown Publishers, Inc.

Library of Congress Cataloging in Publication Data
Main entry under title:
How it works.
1. Technology—Popular works.
T47.H73   600   82-6196
ISBN 0-517-38530-9   AACR2

h g f e d c b a

Printed by L.E.G.O. in Vicenza, Italy

INTRODUCTION

You can't escape modern technology. From television to industrial robots, from controlling model trains to controlling aircraft, it affects all our lives.

This computerized space-age, of artificial hearts on the one hand and sophisticated weapons on the other, is fascinating. But to most of us, it remains mysterious. And even if you do know about some branches of technology, it is unlikely that you know about advances in all the other fields.

This book looks at over 60 examples of technology—from techniques used in everyday life to those that are, quite literally, out of this world. It covers many different aspects: space, military and communications, of course—but also medical science, transport, industry, and the many fascinating ways in which technology helps us enjoy our leisure. It explains them all in straightforward language, and explores how they can be used.

Read this book. You'll find it interesting and absorbing. And you, too, will know *How It Works*.

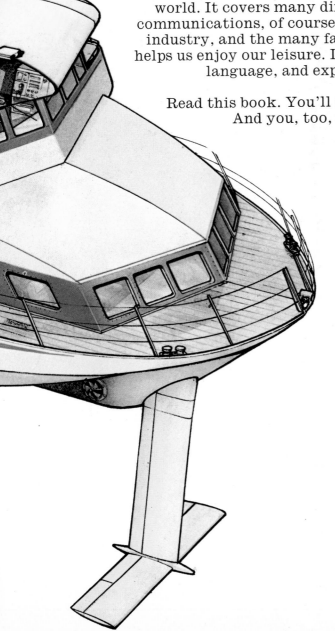

CONTENTS

COMMUNICATIONS

Air Traffic Control 6
Citizens' Band Radio 9
Colour Television 11
Communications Satellites 13
Computers in Printing 16
Computer Memories 19
Electronic News Gathering 22
Television Information Services 25

INDUSTRY

Calculators 28
Cryptosystems 31
Holography 34
Lasers 36
Nuclear Reactor 38
Oil Mining 41
Robots 43
Security Systems 46
Solar Power 49

LEISURE

Animation 52
Aqualung 55
Bowling Alley 57
Hang Gliding 59
Instant Picture Cameras 62
Model Railways 65
Roller Coaster 67
Stereo 70
Video Systems 74

MEDICAL SCIENCE

Body Scanners 77
Endoscope 80
Interferon 83
Kidney Machine 86
Nuclear Medicine 88
Operating Theatre 91
Parts for Hearts 95
Test Tube Baby 99

MILITARY

Armoured Vehicles	102
Bomb Disposal Unit	105
Bullets	108
Cruise Missiles	110
Infantry Weapons	112
MIRVS	115
Submarines	117
Vertical Take-Off Plane	120

SPACE

Ariane	123
Optical Telescope	126
Radio Telescope	129
Space Cameras	132
Space Shuttle	136
Spacelab	142
Viking Mission	146
Weather Satellites	150

TRANSPORT

Airships	153
Electric Vehicles	157
Fire Engines	159
Helicopter	162
High-Speed Trains	165
Hovercraft	168
Hydrofoil	171
Hydrogen Car	174
In-Car Computers	176
Superbikes	179

INDEX

INDEX	182

COMMUNICATIONS

AIR TRAFFIC CONTROL

At any one moment, thousands of people are airborne, travelling around the world at speeds approaching that of sound and beyond. During the past ten years, the number of passengers travelling on scheduled air services world-wide has more than doubled. Journey times have been so reduced that today no two places on Earth are more than two days flying time apart. Some routes, notably in the North Atlantic, are traversed more than 500 times every day throughout the year.

Despite the complexity of marshalling civilian and military aircraft in and out of the world's airports and along a complicated pattern of air routes, air travel remains the safest means of transport in terms of miles travelled. Harmonizing the movements of so many high-speed craft, crisscrossing the globe in greatly overcrowded skyways, is the task of *Air Traffic Control*—an organization of highly skilled, dedicated professionals.

The enforcement of any system of airways depends on a reliable, international network of ground-based navigational aids, located along air routes, which provide the foundation of the air-route structure. Pilots navigate by homing on to the signals of successive radio beacons. At each beacon, they report by radio to ATC, stating their position, height and speed, and obtain permission to continue.

Radar

Radar is considered an indispensable navigational aid at most of the world's airports. It is the major means by which ATC identifies aircraft and keeps track of their height, course, intention and separation. At any one time, an ATC officer might have more than ten aircraft under control—and as many more not far behind. Equipped with a radar display screen, the officer can tell at a glance the aircraft's relative positions, and can instruct the pilots to change course or speed to maintain planned separation between flights.

Two types of radar are in general use: *primary* and *secondary* radar. Primary radar is provided by a revolving aerial, which transmits radio impulses into the atmosphere in all directions. Its main function is to indicate the position of aircraft.

Secondary radar equipment is carried aboard the aircraft, automatically transmitting coded replies to an electronic interrogator on the ground. It is not only more effective than primary radar for determining range and height, but it can also be allotted a digital identification code, which is received on the ground alongside the aircraft's radar response on the controller's display.

Information must be processed before it can be displayed on the radar screen. Each revolution of the radar aerial produces a digital signal containing the height, range and identification of each aircraft detected. The signals are relayed via telephone lines or microwave links to an ATC centre, where they are processed to provide video maps and digital labels containing information about each aircraft being controlled.

Computerization

Some radar stations combine primary and secondary radar data to provide the controller with a single radar display. A further refinement is to process the identification code, together with range, height, altimeter readings and other cockpit information, through a computer to provide up-to-the-minute information about the aircraft.

Many details can be displayed, depending on how the computer has been programmed. They include aircraft speed, rate of climb or descent, destination and controller. All this information constitutes a data block, and it is retained throughout the aircraft's flight. All or part of the data block can be shown on the controller's display.

Essentially, an aircraft's flight comprises three phases: planning, departure/en-route and landing. The flight plan of

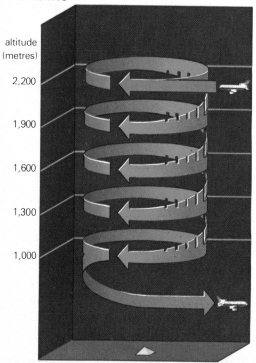

STACKING

altitude (metres)

2,200

1,900

1,600

1,300

1,000

Aircraft arriving during peak times circle at reporting points, or stacking areas, marked by radio beacons while they wait for clearance to land. Maintaining minimum separation distances they descend to the lowest stack from which the final approach to land is made. Modern navigational aids allow aircraft to follow a continuous-descent flight path, which requires minimal engine power and, as a result, significantly reduces noise.

THE INSTRUMENT LANDING SYSTEM (ILS)

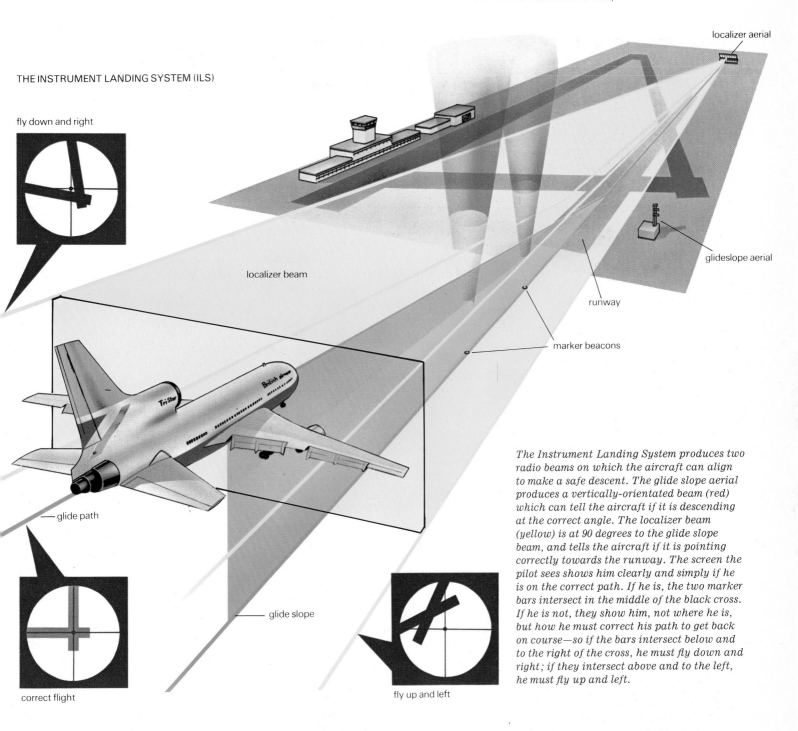

fly down and right

localizer aerial

glideslope aerial

localizer beam

runway

marker beacons

glide path

glide slope

correct flight

fly up and left

The Instrument Landing System produces two radio beams on which the aircraft can align to make a safe descent. The glide slope aerial produces a vertically-orientated beam (red) which can tell the aircraft if it is descending at the correct angle. The localizer beam (yellow) is at 90 degrees to the glide slope beam, and tells the aircraft if it is pointing correctly towards the runway. The screen the pilot sees shows him clearly and simply if he is on the correct path. If he is, the two marker bars intersect in the middle of the black cross. If he is not, they show him, not where he is, but how he must correct his path to get back on course—so if the bars intersect below and to the right of the cross, he must fly down and right; if they intersect above and to the left, he must fly up and left.

a regular flight is usually filed several months in advance so that controllers know beforehand the details of air traffic passing through each sector of airspace. This forward planning is helpful for control.

This flight plan contains such information as the type of aircraft, its call-sign, radar identification code, cruising speed, time of arrival at various points en-route and alternative airports in the event of bad weather at the intended destination.

The information is stored in computers and printed in strips at ATC centres. Strips bearing the details of all aircraft are arranged in geographical order so that each con-

troller can plan safe routes and heights for all aircraft in their control.

Half an hour before a typical flight commences, progress strips are generated by a computer from information contained in the flight plan. When the passengers are on board, the pilot radios the control tower for clearance to start.

After take-off, the pilot activates the secondary radar and the aircraft's identification code appears on the departure controller's display. The departure controller guides the aircraft through its initial climbing stage and transfers control to the en-route controller.

To enable ATC to prevent collisions between aircraft and

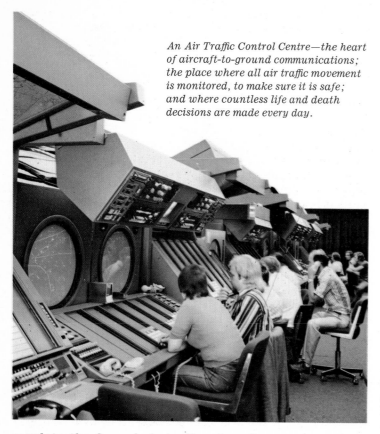

An Air Traffic Control Centre—the heart of aircraft-to-ground communications; the place where all air traffic movement is monitored, to make sure it is safe; and where countless life and death decisions are made every day.

regulate the flow of air traffic, aircraft are separated by minimum distances vertically and horizontally, creating highways in the sky. For example, typically 305m (1000ft) separates one aircraft from another vertically. And aircraft flying in the same 'lane' at the same height must maintain a safe distance apart. A Trident airliner, for example, would be severely buffetted by the turbulence in the wake of a jumbo jet, so a minimum separation of 16km (10 miles) is necessary. A jumbo jet, however, needs to be only 3.2km (2 miles) behind another aeroplane.

A navigational aid used by the pilot while in the air is the *Doppler Very High Frequency Omnidirectional Ranger* (DVOR). This radiates radio signals of specified frequency in all directions. Pilots wishing to fly towards a specific DVOR tune a receiver in the aircraft to its frequency. The receiver indicates the compass bearing that will cause the aircraft to overfly the DVOR. To indicate the distance from a DVOR station, most aircraft are equipped with *Distance Measuring Equipment* (DME) linked to the DVOR receiver in the cockpit.

Preparing to land

When the number of aircraft arriving at an airport is more than the arrival runway can handle, aircraft are held away from the airport in holding or *stacking* positions, where they circle until given clearance to land. Aircraft often arrive at and depart from the world's major airports at the rate of more than one a minute. To handle such peak traffic, a single airport may use up to four separate stacking areas at any one time.

Aircraft maintaining a vertical separation of 305m (1000ft) and following various flight paths join a holding stack at the

upper level, guided by Approach Control. Each aircraft descends through successive levels to the lowest—about 1500m (5000ft). From here, the final approach to land is made and control is passed to a *radar detector*.

Landing procedure

The radar director instructs the pilot when to leave the holding position on a specified heading and commence descent at a set rate and speed. Revised heading, rate of descent and speed instructions are continually issued to pilots to ensure that successive aircraft are safely separated. At the same time, minimum spacings between aircraft ensure maximum utilization of the runway.

The *Instrument Landing System* (ILS) is probably the most exacting of the navigational aids, enabling the pilot to execute the final landing. Essentially, ILS has two aerial systems, each radiating a signal identified by a different audio frequency. Situated at the far end of the runway on the extended centre-line, the localizer aerial radiates a wide, flat beam, which intersects a narrow beam radiated by the glide-slope aerial located alongside the runway. The line of intersection of the two beams establishes the glide path, down which the aircraft flies to land.

In the cockpit, a cross-pointer meter indicates any deviation of the aircraft from the glide path so that course corrections can be made. When ILS is linked to other equipment, such as the altimeter in the aircraft, a totally automatic landing is possible. Automatic landing systems already enable modern jets to land in dense fog down to zero visibility.

ILS provides a single course-line to the runway, but a future development of the system—the *Microwave Landing System* (MLS)—has several advantages over ILS. MLS is not as sensitive to siting techniques. And, most importantly, it can provide multiple approaches to an airport. Aircraft using MLS do not have to follow a straight-in approach along the runway's extended centre-line, so they need less time to make their final turn.

Once the aircraft has been established on ILS, control is transferred to the airport air controller, who is in charge of the landing runway—issuing the final clearance to land and deciding when the aircraft should leave the runway after landing. When the aircraft has cleared the runway, control is transferred to the airport ground controller (also situated in the control tower), who guides the pilot on the surface to the aircraft stand.

Apart from taxi-way lighting, the most advanced aid for ground controller and pilot on the surface is the *Air Surface Movement Indicator* (ASMI)—a radar operating on a very short frequency and a rapidly revolving aerial. It displays a picture of the airport surface. The pattern of runways and manoeuvring areas are clearly visible, as are the aircraft and vehicles moving upon them. ASMI is an invaluable aid at night or in conditions of poor visibility.

Once the aircraft has arrived at its stand area, the airport authority assumes responsibility for processing passengers and baggage. The flight has landed and its associated data—as far as Air Traffic Control is concerned—cease for that particular flight. Over the horizon, the next customers for the team's care and control are crowding in.

CITIZENS' BAND RADIO

For long distance, personal two-way communication, most people use a telephone. But, even if you are lucky enough to have a mobile phone in your car, or a 'bleeper' system in your pocket, a telephone is a restricting device. Two-way radio offers much greater freedom.

To be allowed to use most types of radio system an operator has to pass severe examinations, designed to test his knowledge of the theory and practice of radio communication. The rewards for passing this type of examination are large: you can operate powerful radio equipment and communicate with many other radio amateurs throughout the whole world. But many people just want a simple radio for use over a restricted area—so many countries also license a form of personal two-way radio, often known as *Citizens' band radio* (CB) which permits limited communications, but for which no exam-passing or technical knowledge is needed.

The USA has had a CB radio service for many years, but it has been illegal in Britain. However, a specification for a UK Citizens' Band service was agreed on, and UK CB became legal in November 1981. The specification differs in some technical aspects from that used in, say, the United States. This means that equipment using the US standards will not be properly usable on the legal UK system, and in any case their use will continue to be against the law. Similarly, it will be illegal to use all CB equipment already available in the UK unless modified.

CB radio equipment is deliberately restricted in transmitting power and the frequency bands that it uses, in order to minimize the amount of interference that the system causes both to its own users and to users of other types of radio equipment. A badly aligned or over-powerful CB radio can even cause breakthrough on someone's hi-fi equipment and in some countries interference is a major problem.

Equipment

There are various types, as well as brands and models, of CB equipment. The simplest is the hand portable ('walkie-talkie') which has a built-in power supply and aerial—with this, you can receive and transmit messages anywhere. Because it is portable it is difficult to track down offenders who might, for example, be causing interference. So this class of *transceiver* (combined transmitter and receiver) is usually allowed to transmit only at low power, restricting its range and the damage it can cause.

Slightly more powerful is the mobile station, which is a transceiver for fitting into cars and other vehicles. The most powerful type of equipment, which can be used with a tall fixed aerial, is the fitted base station used in the home.

The range of a CB transmission—the distance over which it can be picked up—depends on the system of transmission used, the type and powerfulness of both the transmitting and receiving equipment, and the location. It might be as little as half a mile or less (under a kilometre) using hand-portable equipment in built-up urban areas, or as far as 20km (12 miles) when transmitting between base stations across open countryside.

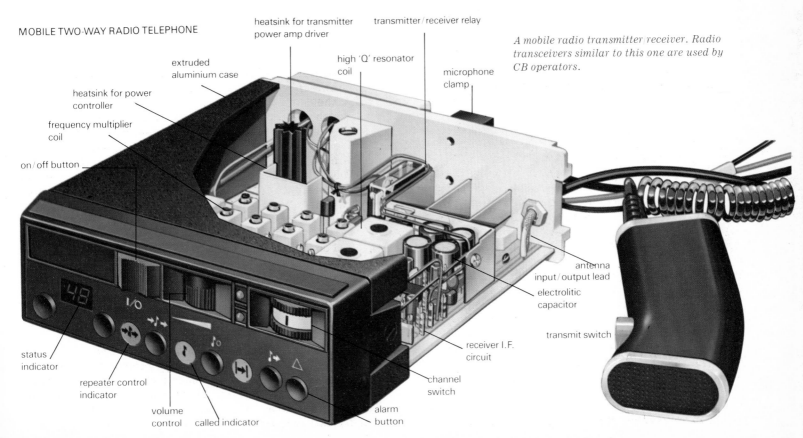

MOBILE TWO-WAY RADIO TELEPHONE

heatsink for transmitter power amp driver

transmitter / receiver relay

A mobile radio transmitter/receiver. Radio transceivers similar to this one are used by CB operators.

extruded aluminium case

high 'Q' resonator coil

microphone clamp

heatsink for power controller

frequency multiplier coil

on/off button

antenna input/output lead

electrolitic capacitor

transmit switch

receiver I.F. circuit

status indicator

channel switch

repeater control indicator

volume control

called indicator

alarm button

A new language has evolved between mobile radio users such as truckers so that transmission times can be greatly reduced.

People in vehicles are big users of CB. They can ask for, and pass on, much useful information about the state of the traffic: which roads are free, and which jammed with traffic; diversions; closed bridges; fog and bad weather and so on. All this flow of information aids traffic flow, to everyone's benefit, and might also reduce accidents. Of course, the information passed on might also be of a more nefarious nature —such as where traffic cops and police speed traps are located. Even this, though, can have its good side—the more reports there are of police, the more likely the drivers are to obey the speed limits, which is, after all, what the police want.

Hand portable transceivers can have a very limited range. But they can be useful for organization in outdoor events, like sports or the village fete, and for keeping in contact with workers on a farm or even a large garden.

The penalty for an unrestricted communication device like CB radio, or indeed, any radio communication, is that messages cannot be private—anyone can listen in to anybody else's discussion. Nor, unlike a telephone, can you 'dial up' any particular person: you can make contact with them only if they happen to be listening to their CB radio at the time you want to call them. The frequency band allocated for CB communications has a large number of different channels (usually 40) so that many pairs of people can hold their own conversations at the same time, even if their transmitter and receivers are within range of each other. However, this increases the difficulty of seeking out a particular person you want to talk to—even if they are maintaining a listening watch on their CB radio, it could be on any one of 40 channels. To reduce this difficulty, one of the channels is set aside for calling purposes—anyone not transmitting listens in on this channel, so callers will know where to find them. Once contact has been made, the two people move to an unoccupied channel. Similarly, another channel is set aside for emergency distress calls—some CB-er somewhere or another will be listening to this channel, which is used for transmitting only genuine distress calls.

Transmission methods

The technology of CB radio is no different from that of any other radio transmission. The transmitter sends out a radio wave—called a *carrier*—at a particular frequency. The audio signal consists of another electrical wave which varies in a way corresponding to what is being said. This audio wave is at a much lower frequency than the carrier wave and is transmitted on top of the carrier wave by making the carrier itself vary in a way corresponding to how audio wave is varying. There are two main ways of varying (*modulating*) the carrier: by varying its size (*amplitude modulation*, or AM) and by varying its frequency (*frequency modulation*, FM). CB systems in most countries use AM, but this is particularly prone to causing interference with other users of that radio frequency. So the system recently approved for CB radio in Britain is to use FM. This causes much less interference, but does call for slightly more complex circuitry.

Radio systems are always described as transmitting or receiving on a particular frequency—CB is usually on 27MHz (megahertz). But in reality, when modulated with an audio signal, a transmission will span a small range of frequencies —its *bandwidth*. The amount of this bandwidth depends on whether the system is AM or FM, and also on the quality demanded. Hi-fi broadcasts, for example, have a bandwidth of about 200kHz (kilohertz—there are 1000kHz in a megahertz) and this limits the number of transmissions that can be squeezed into the frequency band allocated to stereo broadcasting in any area.

CB transmissions require much lower quality, so the bandwidth of each channel can be much less; perhaps as little as 10kHz (which is 0.01MHz). So although, in Britain, CB will be available on 27MHz, the actual transmission frequencies for each channel vary in 0.01MHz steps from 27.60125MHz for channel 1, 27.61125MHz for channel 2, and so on all the way up to 27.99125MHz for channel 40, with each channel spanning a range of frequencies covering 0.01MHz. In the US, the CB service has 40 channels, spanning a slightly different range of frequencies from 26.965MHz to 27.405MHz.

COLOUR TELEVISION

In several Western countries, including Britain and the United States, there are now more television sets than households, and very few homes do not have a television at all. But the technology of television, which we all take for granted, is as complex and fascinating as that of the latest modern electronic inventions.

One of the fascinations of television is the number of complex problems that have to be solved before a working system can be developed. How can a visual scene be captured electronically? How can the electrical impulses be transmitted? How can they then be re-converted into a visual image? And how can the system cope with the colour information in a scene?

A useful place to start answering these questions is at the television receiver. The device that is used to re-convert electrical images into visual images is the *cathode ray tube* —just about the last valve (tube) left in a modern TV set. The broad front of a cathode ray tube (crt) forms the screen of a TV set, and the inside surface of this screen is coated with a phosphor which glows momentarily whenever a ray of electrons hits it. Such a beam of electrons is produced by a 'gun' at the other end of the crt. The brightness of the glow

on the screen depends on the strength of the beam of electrons, and this is easy to control electrically by varying the voltage which produces the beam. So already we have a way of producing visual impulses in terms of varying lightness and darkness from electrical impulses.

The next problem is to control which parts of the screen are to be illuminated to each particular level, so that a recognizable picture can be displayed. At first sight, it might seem that the way to do this is to divide the screen into thousands of tiny areas and provide each one with its own electron beam—as long as the areas were small enough, the picture would appear as a continuous whole to the human eye. However, it would be impossible to build and control a crt that used thousands of separate beams. The answer is to use just one beam that moves in lines across the screen from one side to the other, in much the same way as people read the lines of print in a book: this process is known as *scanning*. As the beam scans the screen, its intensity can be continually varied, so each minute length of each scanning line can be of a different brightness. As long as the lines on the screen are close enough together, then a clear picture will be displayed.

Of course, it takes a certain amount of time for the beam to complete a scan of the screen, and to return to the top and start again—and it might be thought that this prevents the idea of scanning being practical for instant, moving pic-

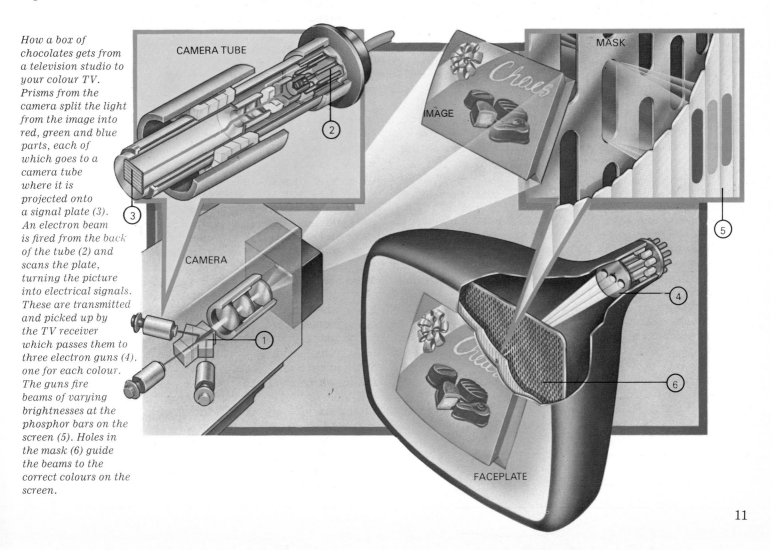

How a box of chocolates gets from a television studio to your colour TV. Prisms from the camera split the light from the image into red, green and blue parts, each of which goes to a camera tube where it is projected onto a signal plate (3). An electron beam is fired from the back of the tube (2) and scans the plate, turning the picture into electrical signals. These are transmitted and picked up by the TV receiver which passes them to three electron guns (4), one for each colour. The guns fire beams of varying brightnesses at the phosphor bars on the screen (5). Holes in the mask (6) guide the beams to the correct colours on the screen.

CAMERA TUBE

MASK

IMAGE

CAMERA

FACEPLATE

tures. This is not so, mainly because the human eye retains an image it sees for a short period of time after the image has changed. You can see this persistence for yourself, if you flap your hand very quickly up and down—the image of your hand at the top of its flap is still there, even after your hand has moved down. So if a scan can be completed and repeated within the time that an image is retained, the eye will be fooled into thinking that the image is there all the time. Similarly, even if the picture consists of a moving scene, as long as one scan follows another rapidly enough, the amount of movement between the two pictures will be too small to appear 'jerky' to the eye. Modern television completes 50 (in Europe) or 60 (in the United States) scans every second—enough to fool the eye.

Strictly, each scan forms only half a picture, as the electron beam traces out the odd-numbered lines of the screen on one scan, and the even numbered on the next scan—this *double-interlaced scanning* makes transmission technically more economical.

The number of complete lines in a television picture depends on the country and varies in different countries—most of Europe uses 625 lines, though an 819 line system has been used in France, and the USA uses 525 lines. The more lines in a picture, the sharper and clearer it can look, though there is little difference in practice between a 625 line and a 525 line picture.

We now have a television receiver capable of displaying reasonably realistic pictures, in response to a changing electrical impulse. Next we need a television camera that will view a scene and produce corresponding electrical impulses that will be accepted by the receiver.

TV cameras

The technique used in the camera is much the same as that of the cathode ray tube in the receiver, but in reverse. Light from the scene to be transmitted enters the camera through a lens which focuses it on to a light sensitive screen at one end of a tube. An electron gun at the other end of the tube fires a beam of electrons at the screen using the same scanning motion as in the receiver. As the beam moves over the screen it produces, in various ways according to the type of camera, a stream of electrical impulses which vary in electrical intensity according to the amount of light falling on the part of the screen that the beam is hitting.

All that is left to do now is to provide some way of transmitting the electrical impulses produced by the camera in such a way that they will force the electron beam in the receiver's crt to trace out the same visual image that the camera sees.

Transmission

The actual transmission of the impulses is not too difficult—the main problem lies in making sure that the camera's electron beam and that in the receiver's crt do their scanning in step with each other—if they do not then the picture will break up. To ensure this, the transmission contains a number of *synchronizing pulses*. One pulse is sent at the end of every line, to tell the receiver's electron beam to move back and start a new line (during this *flyback*, the beam is blanked out, so that it does not show on the screen). More pulses are

sent when the beam reaches the bottom of the screen, and to allow time for these to be transmitted, not all the lines in a system are used for the picture. However, only part of the time left by this *field blanking gap* is in fact used by synchronizing pulse and the remaining gap has been found a useful place to send test signals for engineers, and an information service (see page 25).

Colouring the picture

Producing colour transmission and reception is the final problem to tackle. When colour television was first seriously studied, it was realized that not everyone would want to buy a colour receiver. Nor would it have been feasible to provide totally separate colour and monochrome (black and white) services. So a way had to be found of making a single service *compatible*—that is, so that colour transmissions would be receivable (in black and white) on a monochrome set. The solution is to split the transmission into two parts, the major part carrying all the black and white (*luminance*) information, including the shapes and outlines and all the light and dark shading and the other part carrying just the colour information (*chrominance*) which is coded and added to the luminance signal for transmission. There are a number of ways of electronically carrying out the encoding (and some variations, too, within the ways) used by different countries.

The American NTSC (standing for National Television Systems Committee; the body that laid down the standard) was the first system. PAL (Phase Alternate Line) was introduced a little later and has some additional refinements which give it better picture quality than NTSC; it is used by Britain and many west European countries. SECAM (Système en Couleurs à Mémoire) was developed by the French and differs more than the other two systems; it needs a simpler form of receiver. It is used in France, and by some east European countries.

It might be thought that transmitting and receiving colour pictures is an impossibly difficult task, even when the problem is reduced to simply transmitting colour information—after all, there is an infinite range of colours. But, again, there is a simple short cut—all colours can be successfully produced by mixing together various amounts of just three colours: red, green and blue. So a complete colour transmission consists of luminance information, plus a signal giving the proportions of red/green/blue at each moment in the picture.

At the receiver, a colour television picture tube works on the same principles as a monochrome one, but is more complex. The phosphor screen is divided up into thousands of tiny dots or bars, grouped in threes. In each group, one dot will glow red, one blue and the third green. The dots are so tiny and close together that the overall effect is that each group can glow with any possible colour, depending on the relative brightness of the three dots.

There are three electron beams, one of each to react to the red, green and blue parts of the colour signal; a special mask in the front behind the screen ensures that the 'red' beam hits only the red dots on the screen and so on. In this way, the colour signal makes the coloured phosphor dots glow to the right intensities, producing an image with the same colours and hues as the original.

COMMUNICATIONS SATELLITES

Communications satellites are used more and more the world over. Circling the Earth in 'stationary orbits', they handle as many as 13,000 telephone calls at any one time and relay live colour television pictures from continent to continent. But satellites are expensive, involving a mass of equipment and a costly launching process. Radio broadcasts can reach anywhere in the world without using satellites. And many telephone conversations already travel along a network of existing network of undersea cables. So why are satellites necessary at all?

It is certainly true that some radio broadcasts can travel round the world without the aid of satellites. But this is only because radio broadcasts are usually transmitted at relatively low frequencies. If the frequency is very low, the transmissions simply bend round the earth, following the curvature of the earth. Slightly higher frequency transmissions cannot do this, but can make use of the electrical properties of the *ionosphere*—an upper layer of the atmosphere surrounding the earth. As these frequencies hit the ionosphere, they are reflected back to ground again, which vastly increases the distances over which the transmissions can travel.

Television broadcasts, however, have to be transmitted on very much higher frequencies: these transmissions not only travel in very straight lines, but are also not reflected by the ionosphere. Consequently, normal television transmission is limited to a couple of hundred kilometres. A satellite, however, can be used to bounce a signal transmitted to it

A cutaway view of OTS 2. The satellite, one of a series, is built in modular form to give maximum flexibility. The central service module remains much the same for all missions, but the communications modules can be changed on subsequent launches. Satellites are growing in size as they get more powerful.

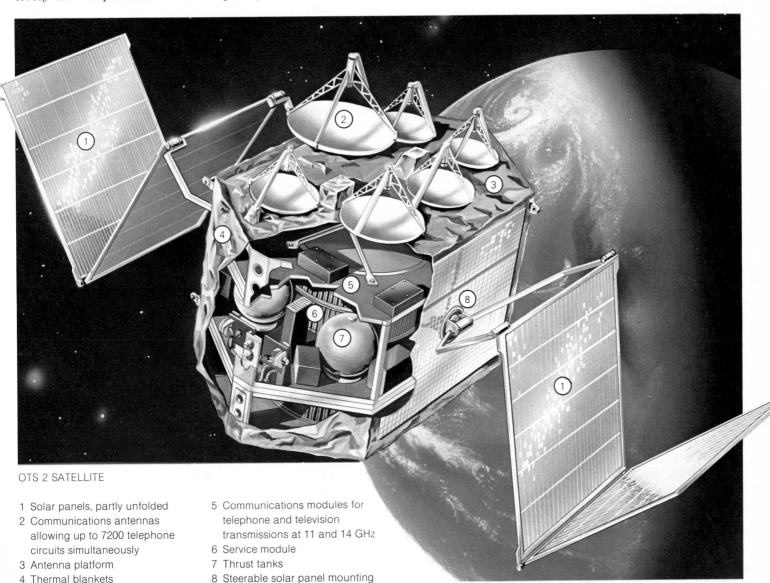

OTS 2 SATELLITE

1 Solar panels, partly unfolded
2 Communications antennas allowing up to 7200 telephone circuits simultaneously
3 Antenna platform
4 Thermal blankets
5 Communications modules for telephone and television transmissions at 11 and 14 GHz
6 Service module
7 Thrust tanks
8 Steerable solar panel mounting

back to earth in the same way as the ionosphere does, making possible television transmissions world-wide. And because a satellite is man-made, it can be designed to cope with broadcasts of any frequency, unlike the restrictive and fickle ionosphere.

With telephone conversations and all the other signals that are sent along telephone cables, such as facsimiles and computer data, there are two main problems. The first is that cables are fragile and expensive to lay. The second is the sheer quantity of today's communications. Every individual telephone line requires a pair of wires and undersea cables contain many such wires, but their numbers are still insufficient to cope with all the material that comes flooding through.

By coding the information on to a radio broadcast, though, much more material can be handled simultaneously. But these transmissions are not easy to send over long cables—they are best broadcast over the air, and a satellite is a good vehicle to do this world wide. Moreover, once you have decided to code telephone messages and so on onto a radio broadcast, it makes sense to go for the highest transmission frequency you can get—the higher the frequency, the more signals then can be carried at the same time. Satellite transmissions are at a very high frequency—they can handle up to 13,000 telephone lines simultaneously. (Colour TV broadcasts use more frequency space than a telephone conversation, so far fewer—only about 40—can be handled at the same time.)

Intelset

The commercial Intelset communications network was set up during the late 1960s and has been maintained and improved since. A satellite is expected to last for seven years though some have worked for ten. After this a new satellite is sent up to take its place—each replacement satellite tends to be bigger and more complex than its predecessors. The launchings are carried out by the USA, but any country can hire time on the network.

A modern communications satellite looks rather like a plant in a pot. The 'leaves' of the plant are the dish reflectors of the transmitting and receiving antennas. These concentrate the incoming signal and focus the relayed signal into either a narrow or a wide beam, as required. Fine adjustments are made to the attitude of the satellite by little rocket motors fuelled by hydrazine and fired in response to a signal from Earth. The satellite carries enough fuel to last its lifetime—indeed, in future the lifetime of a satellite might be limited by the amount of fuel it can carry.

Electrical power comes from solar cells covering the walls of the cylindrical body. Only a few hundred watts are generated, but this is enough to power both the spin motor and all the electronic equipment, though more power (which would allow more and better communications) is always sought after. Batteries provide a power reserve for when the satellite is in the Earth's shadow.

The electronic equipment is complex but compact, and makes much use of miniature integrated circuit 'chips'. The original motive for designing these tiny devices was the demands of the space programme, with its insistence that everything should be as small and light as possible. But ironically the latest chips are now too small to be used in space.

The individual components of an advanced chip are so tiny that cosmic rays—streams of high energy particles—could cause damage. If a particle scored a direct hit on a sensitive spot it would change the electric charge on it so that the chip garbled a transmission or forgot a vital instruction in its 'memory'. The slightly larger, not-quite-latest model, chips are tough enough to stand up to cosmic rays and also to resist the vibration and stress of launch from Earth.

Satellites are carried into orbit by three-stage rockets. The first two stages of the rocket carry it into a low orbit

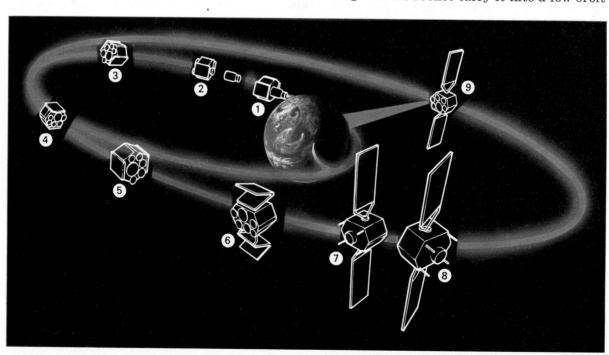

OTS 2 LAUNCH SEQUENCE

1 Thor Delta third stage burn
2 Satellite separation
3 Satellite transfer orbit
4 Injection into near geosynchronous orbit
5 De-spin
6 Solar arrays deployed
7 Attitude acquisition
8 Ground station contact
9 Satellite normal mode

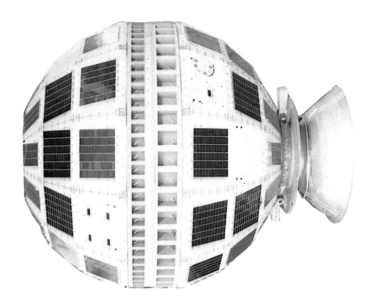

The satellite which relayed the world's first live transatlantic television pictures, Telstar 2. Around the middle of the satellite are the microwave antennas which receive and re-transmit the signals from earth.

One of the antennas at Goonhilly in Cornwall, England, near the spot from which Marconi made the first transatlantic broadcast.

from which the third stage boosts it into a high, elliptical orbit before separating from the satellite. At the apogee—the top of the ellipse—the satellite's own 'apogee motor' is fired to bring it into a circular, synchronous orbit. Final adjustments have to be made by short firings of the apogee and attitude control motors before the satellite is positioned in exactly the right spot.

Positioning a communications satellite is, in fact, crucial to its operation—it must circle the earth at a height of exactly 35,880km (about 22,400 miles), and lie above the equator. Why? The time it takes for a satellite to orbit the earth depends on the height of its orbit—at 35,880km, it takes exactly the same time to circle the earth that the earth takes to rotate about its own axis. So the satellite will appear to be fixed at the same point above the ground all the time, which makes it easy to send and receive transmissions. Such an orbit is called a *synchronous* or *geostationary* orbit.

The first TV broadcast by satellite between the USA and Europe (in 1962 via Telstar 2) lasted only a few minutes because it could be maintained only while the satellite was 'in sight' from both New York and London. Telstar, in a low orbit, rushed across the sky and soon disappeared over the horizon. During this time it had to be tracked by moving antennas transmitting and receiving the broadcast. A satellite in synchronous orbit, however, is always in the same place in relation to the Earth and so antennas can be stationary. Three such satellites spaced equally round the Equator will give worldwide coverage, except for the extreme polar regions.

Ground stations

Back on Earth, *ground stations* provide communications between the satellites and their users. There are over 100 of these involved in the commercial satellite network alone. The largest of them, which deal with colour TV broadcasts as well as telephone traffic, have a transmitting power of several kilowatts. The more power is given to transmitters on the ground and the more sensitive the ground receivers, the less power is needed to run the satellite.

Communications is a growth industry. More and more countries are demanding time on commercial satellites, and this includes many Third World countries. Often these do not have an established network of ordinary TV transmitters, and it would be convenient for them if instead of building these they could beam satellite transmissions directly down to individual receivers. This could in theory be done using ground receiving dish antennas about 1.5m (5ft) in diameter, extra-powerful receiving sets and advanced digital coding techniques. In the near future, however, individual satellite reception is likely to be provided by *direct broadcasting satellites*—technically little different from communications satellites, but providing a signal over a small part of the world.

Most of the cost of a communication satellite goes in launching it. It is hoped that the reusable US Space Shuttle (see pages 136 to 141) will usefully lower launch costs.

Another possibility raised by the Shuttle is that it may be possible to visit and repair satellites in orbit, thus prolonging their usefulness. Existing satellites are not designed to be repaired in space, but future ones might be built with easily accessible plug-in component boards rather like a modern TV set. Astronaut engineers could then go up with replacement boards, solar cells, batteries, hydrazine and so on.

The future will eventually bring another problem. Although there is plenty of room on the single, synchronous orbit occupied by nearly all communication satellites at the moment, it is not desirable that satellites should be too close to each other. Already a couple of thousand large pieces of dead, non-working 'space junk' are circling the planet, along with countless smaller items. It is hoped that the Shuttle will be able to clean up some of the rubbish: the first space-age garbage truck.

COMPUTERS IN PRINTING

Modern business, leisure and pleasure all thrive on the printed word. Whether you want to know the progress of the stock market or the latest way to customize your car, or you simply want to relax and be entertained, there is a newspaper, magazine or book to help you. And, despite electronic revolutions, such as teletext and viewdata, there are still many people who will want their words printed on ordinary paper for many years to come.

The problem with printing words on good old-fashioned paper, rather than making use of electronic display units, as teletext and viewdata do (see page 25), is that, until recently, it has been a very expensive business. The main problem has been, not so much the actual printing of the newspaper, or whatever—automation has made continuing progress over the years—as setting the type to be printed in the first place.

In the beginning
The oldest methods date from the very birth of printing, yet are still used today for small printing jobs. Each word to be printed is put together by the printer, working by hand, from individual letters of type. Type consists of small blocks of lead alloy, with the letter itself forming a raised surface at one end. If an inked roller is passed over the block, only the raised portion will receive ink, and if the type is subsequently pressed onto a piece of paper, the form of the letter will appear on it. Setting individual letters like this is obviously very time-consuming, especially for a big job—this book, for example, contains well over half a million individual letters and spaces. An additional irritation is that type has to be set as a mirror image, in order to appear the right way round when printed. So the printer has to be skilled in reading lines that look like:

The quick brown fox jumped over the lazy dog.
(If you cannot understand this line, hold the book in front of a mirror.)

Mechanical machines
The only major advance in this end of printing in over 400 years was the introduction of mechanical machines to do the job of actually putting the right pieces of type in the right places and in lines. Now the printer could sit at a keyboard, much like a typewriter, and type in the words to be printed. This system, though, is still very slow and not tolerant of mistakes—corrections have to be made by hand. It is important not to underestimate the need for corrections. Even if the printer does a first-class job, and the author on rereading his article does not want to make alterations, it is often necessary to change the length of an article when first printed so that it fits into the space allocated for it. For example, this article must finish just above the picture on

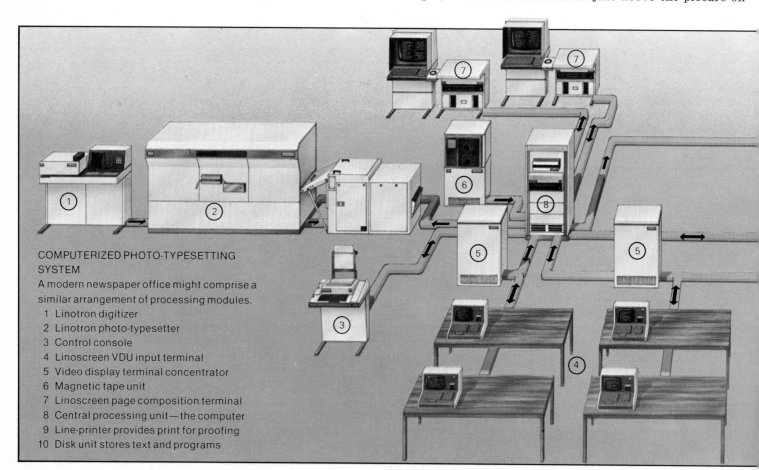

COMPUTERIZED PHOTO-TYPESETTING SYSTEM
A modern newspaper office might comprise a similar arrangement of processing modules.
  1 Linotron digitizer
  2 Linotron photo-typesetter
  3 Control console
  4 Linoscreen VDU input terminal
  5 Video display terminal concentrator
  6 Magnetic tape unit
  7 Linoscreen page composition terminal
  8 Central processing unit—the computer
  9 Line-printer provides print for proofing
 10 Disk unit stores text and programs

page 18. If, when the typewritten copy is first printed, it is too long, then parts will have to be re-written (and therefore re-set) to shorten it without losing essential information. So a modern system of typesetting machinery would have to do two jobs easily and quickly: it must allow alteration of the original text, and it must set up the text in type form. Modern electronic technology, incorporating computers and even lasers (see page 36) can do this. But electronic typesetting owes some debt to advances in printing; it is now no longer necessary for type to be set in the form of raised letters: printing machines can work from an actual copy of the text typeset on special paper or photographic film—in roughly the same way as an ordinary photocopier does the same task.

Phototypesetting

In a modern typesetting machine, the letters are formed, still one by one, as images directly onto special paper—this paper then has the exact appearance of that part of the final printed and published book. There are various ways of creating the letter images. In the original phototypesetting machines, each letter is held on a photographic slide, and a large number of slides are mounted together in a disc. When the printer presses a key on the keyboard, the disc spins round·to bring the correct slide in line with a projector system which 'photographs' the image of the letter in the correct place onto a piece of light-sensitive paper: letter by letter, the text is built up in the same way. A phototypesetter

has far fewer moving parts than a machine for setting in metal type so it is quicker and more reliable, and it need store only one example of each type of letter to be printed rather than hundreds. And the letter images can be reduced or enlarged by the projector's optical system, so different sizes of type can easily be accommodated.

But there are disadvantages: the system is still mechanically based, so the number of different type styles and special characters that can be stored in the machine is rather limited; if more type styles are called for then the operator has to stop the photosetter and remove and change a disc of slides, which is time-wasting. So modern photosetters rely on a combination of computer memories and cathode ray tubes (CRT) as in a television screen to replace the photographic slides and projector systems. The shape of each letter is held in a computer memory, and when it is wanted, it is drawn on the CRT screen by a scanning process, similar to that used in television (see page 11) but with the scanning lines very much closer together. Light sensitive paper is held firmly up against the screen, so the image of the scanned letter is photographed directly onto it. With this sort of typesetter, there are no moving parts (except for moving the printing paper) and the range of type styles that can be stored is limited only by the size of the computer memory. The latest photosetters use a laser beam to 'burn' the image of the letter directly onto the printing paper. As with CRT types, the laser beam is directed in its scanning and told when to turn on and off by computer.

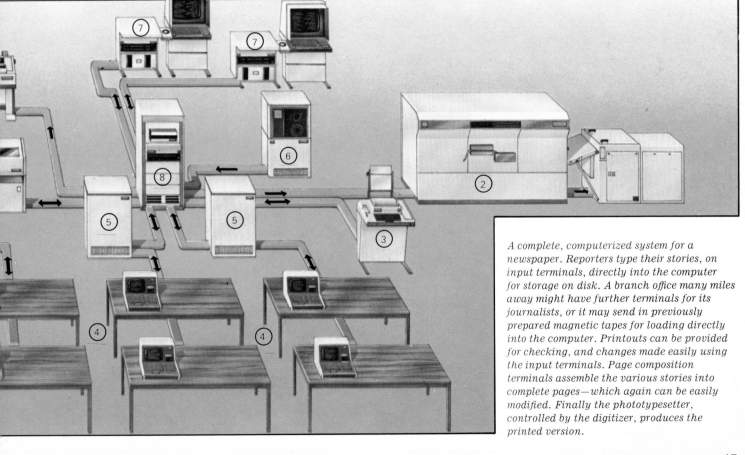

A complete, computerized system for a newspaper. Reporters type their stories, on input terminals, directly into the computer for storage on disk. A branch office many miles away might have further terminals for its journalists, or it may send in previously prepared magnetic tapes for loading directly into the computer. Printouts can be provided for checking, and changes made easily using the input terminals. Page composition terminals assemble the various stories into complete pages—which again can be easily modified. Finally the phototypesetter, controlled by the digitizer, produces the printed version.

In a sense, the output from a photosetter is more, not less, difficult to correct than the output from a metal-setting machine, since individual letters and words cannot be altered. Usually, a whole paragraph will have to be re-typed, and the changed text carefully stuck into position over the wrong copy. Re-typing a lot of copy (most of which will be correct) is very time wasting, and there is always the possibility of new mistakes occuring. The answer is to provide a way of changing the text electronically, and sending it for photosetting only when it is accurate.

Systems for doing this, which are an integral part of photosetters, go by a variety of names such as *text capture and editors*, *front-end systems*, and so on. In the front-end system, the text is typed on a keyboard, and each letter and space is stored in a computer-type memory. The text can be displayed on a VDU and, with a few simple commands entered from the keyboard, the printer can correct errors and omissions, erase any part of the text, and change the order of words or even whole paragraphs. When he is satisfied, the text can be stored permanently on magnetic discs and, using an ordinary computer printer, a typed copy can be given to the customer so that he can add his changes, which again can be made easily and electronically. More importantly, the system will store all the printing commands that the photosetter will later use when producing the printed version of the text. For example, the printer will want to specify the sizes and styles of types, the lengths of lines, whether the copy is to be set with a smooth right-hand edge (as this copy is) or ragged (as the captions are). Where a whole word will not fit on the end of the line, the computer will hyphenate it at a sensible place, for example between syllables.

So the text produced by the computer printer shows as closely as possible what the final printed version will look like—and authors are rapidly losing all their traditional excuses for making extravagant and expensive changes at the final printing stages. The text can be shortened if necessary, and another computer printout produced to show whether enough, or too much, cutting has been done. The effect on the length of an article of changing the size or style of type can easily be seen. Just a few commands need be typed in to alter the whole format and style of the text, and a new printout can be produced easily and cheaply.

This sort of flexibility and speed is particularly in demand for setting newspapers, where time and space are always in great demand, and corrections must be allowed up to the last possible moment, if the paper is to be accurate and topical. Modern setting equipment can do all this and more—it can display whole pages of a newspaper, showing what happens when copy is changed or moved from one column to another. Other systems can even help to design advertisements.

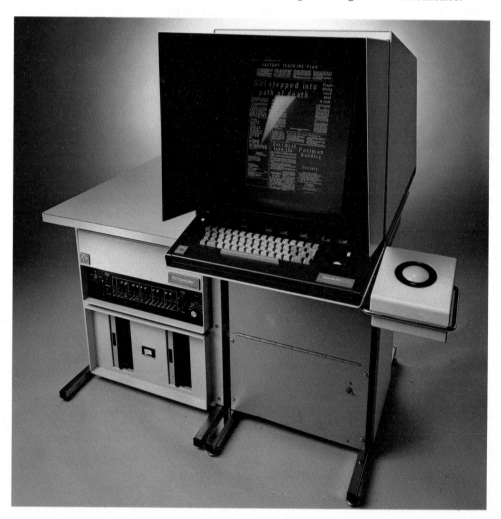

This Linoscreen VDU terminal is displaying a page from a local newspaper, showing exactly what the page will look like when printed—with all the headlines and the text copy in their correct sizes of type, and space left for any pictures. At this stage, the page does not exist printed on paper in this form— all the words are stored in a computer memory, together with typesetting printing commands detailing how big all the text and headlines are to be set, and where they are to be placed on the page. If changes are necessary—for example, if fresh information becomes available on a news story—they can be made easily and quickly by typing in new copy and commands at the keyboard; the computer takes care of the rest.

COMPUTER MEMORIES

The fickle nature of Man's own memory, with its annoying capacity to forget or even alter facts that have apparently been remembered, is a weakness he has long struggled to overcome. Through much of 'recorded history', pen and paper have been the most useful and durable alternative to 'remembering'. Even now, the sum total of Man's knowledge through the centuries is stored on paper. But the advent of electronic memories has opened an astonishing new chapter in the complex task of keeping track of knowledge.

Computers are designed to do two main jobs. One is to carry out very simple calculations, such as adding two numbers together, or comparing two numbers and deciding which is the larger. The range of these calculations is very limited—computers cannot even multiply or divide directly! The second job they do is to store numbers.

The ability to store and remember information is absolutely vital if we want the computer to carry out more than the very basic of tasks. Then, when a computer has carried out a calculation, it can store the result, carry out another calculation, and so on. By carrying out simple tasks, one after the other, and storing and retrieving the results at will, the computer can rapidly carry out complex calculations.

The type of memory a computer needs to do this sort of job is called a *random access memory* (RAM). Random access means that the computer can find the number it next needs from among the many thousands that it has stored directly, without having to first sort through all the stored numbers. To do this, the computer must keep track of where in its memory each number is stored, so every storage location has its own unique label, graphically called its *address*, and the process of directing a number to a particular location or picking a number out of a location is called addressing.

Computer numbers

A computer could be designed to store numbers and allocate addresses using the traditional decimal system where numbers are represented by ten different symbols (0, 1, 2 and so on up to 9). But this would be very cumbersome, and computers find it much easier to use a number system which has just two symbols, which the computer represents as 'on' and 'off' states within its electronic circuits. Humans allocate the symbols '0' and '1' to these states, and the system is known as the *binary notation*. Numbers represented in binary notation require more digits than those in decimal notation. For example, to represent the two digit decimal number 13 requires four binary digits or 'bits'—1101.

This number can be presented to the memory one digit at a time, followed, after a time interval, by another train of digits corresponding to another number, in what is known as *serial addressing*. Alternatively, it can be presented all at once along a number of parallel wires to the memory, each of which carries one bit. This is known as *parallel addressing*. Parallel inputs generally accept 4, 8, or 16 bits of information each in the computer. (If a binary number requires less bits than this the other wires carry zeros.)

Whichever way the number is presented, each 0 or 1 has to be stored in its own electronic circuit using at least one transistor. The memory simply consists of row after row and column after column of separate transistor circuits, these days all contained on a tiny silicon chip, each capable of storing just one binary digit. The chip must also carry circuitry needed for checking, sorting, sending, seeking and finding information entering and leaving the memory. Such memories have now reached the point where over 65,000 bits can be stored on a chip and the manufacturers of these memories could make them even smaller if the size of the physical connecting wires and tabs could be reduced.

There are two types of semiconductor RAM: *static* and *dynamic*. The dynamic RAM loses its memory after a time and so has to be 'refreshed' regularly. This means that extra circuits are needed to refresh the memory, but each memory cell or storage unit on the chip is smaller than that of a static cell, which requires at least two transistors to store one bit of information.

In the static memory, the signal is stored by a device called a transistor flip-flop, a switch that can be only on or off, so there is no need to keep a cell charged up, but the static memory may consume more power because one or other of

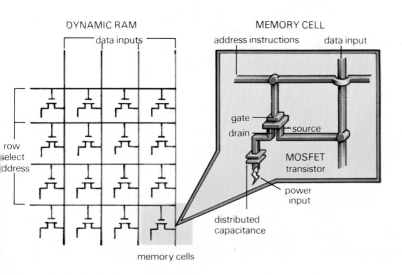

DYNAMIC RAM
data inputs
row select address
memory cells

MEMORY CELL
address instructions
data input
gate
drain
source
MOSFET transistor
power input
distributed capacitance

A section of dynamic RAM, using fets, showing how the memory cells are connected in rows and columns. Data can enter and travel down a particular column, but can enter only the fet whose gate has been turned on (this is done by additional circuitry). The number of the row selected is called the 'row-address information'. In serial addressing, the date are accompanied by 'column-select information' so that circuitry can direct data into the right column (in the same way as it will then direct it into the correct row). In parallel addressing, there is no need for this additional circuitry, because the whole of a binary number is presented, one digit to each column, at a time. To read a number out of the memory, a request for the number is passed to the correct row address. The gates along that row are turned on a and if there is a 'one'—meaning a charged capacitor along any one of the data lines in that row—the pulse from the capacitor will be passed to the data line. If a cell along that row has a 'zero' there will be no effect on the data line.

the two transistors in each cell is conducting the whole time. The conduction also produces heat, which needs to be dissipated and tends to add to the space needed for a cell, which is already relatively large. Dynamic RAMs have now reached a capacity of 64,000 cells per chip, compared with a quarter of this for the latest static RAMs.

The type of transistor most commonly used for memory storage is the Metal Oxide Silicon or Metal Oxide Semiconductor (MOS) transistor. This is a *field-effect transistor* (fet), of which thousands can be fabricated onto a square centimetre of silicon.

In a memory, the fet is used as a switch which can be turned on by sending a signal to one of its terminals called, appropriately, a gate. Once in the 'on' condition, current may pass through the device to charge up the memory storage capacitor or 'cell' associated with the particular fet. This capacitor can take up either a state of charge, seen as a binary 'one', or a state of discharge, seen as binary 'zero'. So, by this technique, binary information may be stored, one digit to each fet.

The fets have to be arranged in a way that will make addressing (choosing the right fet for each binary digit) easy. This is done by connecting them in rows and columns. See the diagram below.

Using techniques like these, large amounts of storage can be built up on a single chip. Refreshing (the type of memory just described is a dynamic RAM) is carried out by applying a 'topping up' signal in turn to each column of cells. This is done about 500 times a second.

Polysilicon

Not all RAMs use MOS technology, even though the most rapid developments in memories have taken place in the last ten years using MOS. *Bipolar* memories (using a different type of transistor) are also used, but they tend to be designed

in the same way as MOS static memory, with pairs of flip-flops and a consequent increase in the area needed per memory cell. Bipolar memories, unlike their MOS static equivalents, can be much faster than dynamic memories, though MOS dynamic memories remain the cheapest. But MOS memories may need more than one power supply, and they have to be supplied with circuitry that will allow them to work with the very fast logic of the central processor, which is likely to use bipolar circuits.

A variant of the MOS static RAM is the Complementary MOS or CMOS RAM. In this the load resistors of the crossed transistor pair in a flip-flop are replaced by transistors which are switched in the opposite way to the actual switching transistors. The resistance presented by these two load transistors is, in theory, infinite. So no current flows and no power is dissipated. These consume far less power than other types so they are used extensively in space research. Unfortunately, however, they tend to be far slower in operation.

One disadvantage of all semiconductor RAMs is that they are *volatile*—if power to the memory is removed, all the cells lose their information. This necessitates an elaborate system of back-up power supplies for use in the event of a power failure, or the transfer of important information immediately to some form of permanent storage, such as magnetic tape or disk.

However, there are types of semiconductor memory that are non-volatile. These tend to be used to store programs indefinitely, perhaps for a TV game or to direct operations in a computer, or to instruct a calculator on how to perform a mathematical process. The computer cannot write data into such memories, so they are called *Read Only Memories* (ROM). However, ROMs are still random access for reading —the computer can reach the data it wants without having first to sort through all the other data.

Essentially, such memories consist of arrays of diodes in a matrix and programming is achieved by removing unwanted diodes, by fusing them, through the application of a high voltage. The resultant lack of connection between lines may then be used to indicate a logic state of 'one' or 'zero', according to the way the ROM is connected. The simplicity of a ROM is the reason for its low cost and high information density.

Early ROMs were incapable of alteration once they had been programmed. But soon there appeared devices that could be 'wiped clean' by exposing their internal diode arrays to ultra-violet light, through a small window in the top of the device. These are called Eraseable Programmable Read Only Memories (EPROMs), as opposed to the programmable ROMs (PROMs).

The EPROM uses the characteristics of certain semiconductors whereby, without destroying any part of the whole array, each part of the matrix could be programmed by applying something like 30 volts to it. If, however, ultra-violet light were shone over the whole array, this program could be completely wiped out. Such EPROMs have now reached densities of 64,000 bits, and work is well advanced on 128,000 and 256,000 bit versions of these devices.

Another variation of the ROM is the electrically alterable ROM or EAROM. With these it is not necessary to wipe the whole memory, since application of larger-than-normal voltages will individually re-program cells. Thus one memory cell can be changed at a time without the need to re-program the whole matrix—which is fast approaching the ideal of a true, yet non-volatile RAM.

In the future

There are many other types of storage for digital data, many of which, although promising, have yet to make their impact on the memory market. Among these the most important examples are *magnetic bubble* memories and *charge-coupled device* (CCD) memories.

In both these memories the problem is access time, since they are not truly random access devices—the stored data are repeated endlessly in sequence, and the computer has to wait until the piece of data it wants 'comes round'. Consequently, such memories are likely to become substitutes for disk storage.

In the case of the magnetic bubble, information is stored by using the properties of certain crystals which, under the influence of a magnetic field, store what look like bubbles of magnetic polarity. Alignment of the bubbles determines whether a 'one' or 'zero' is stored. An alternative is to store either a bubble or no-bubble, signifying a 'one' or a 'zero'.

Such devices are capable of storing vast amounts of information, but the information has to be continually circulated for the device to function properly.

As a result, a special device is needed to 'write' information into the store, rather like the recording head of an audio tape machine, and another is needed to 'read' it out. The latter can read the information out only when it is available during circulation. This makes access time rather long, but, like other magnetic devices, bubbles do have the advantage that their memory is non-volatile and packing density is something like ten times that of a semi-conductor memory. Such memories are now beginning to be used by companies like IBM, and some of the Japanese computer makers.

Charge-coupled memory devices use a semiconductor technique—a very long fet with a number of taps or electrodes along its length. Information is stored in a long train as amounts of electrical charge, each new piece of data being added to the end of the train.

A clock signal has to be applied to keep the information ticking around the device, and the access time to any bit of information is higher than with other comparable techniques, but cost is said to be low. As a readily useable technique of memory storage this once-promising method has yet to prove itself suitable for widespread use.

Below: An Intel bubble memory with one megabit storage.

Left: The characteristic appearance of an EPROM (erasable programmable read-only memory). The chip is visible through the window needed to erase the memory by exposing it to ultra-violet light.

ELECTRONIC NEWS GATHERING

Electronic news gathering—also known as electronic journalism—is a technique that replaces TV news film with electronic pictures. It enables newsmen to cover events and get their pictures on to viewers' screens more swiftly than ever before possible.

The technique evolved in the US, where there is fierce competition to be first with the local TV news. Stations have heavily publicized their own investment in electronic news gathering (ENG) with phrases like 'Eyewitness News' and 'Instantcam'.

Live coverage of events—from sporting occasions to embassy sieges—has long been possible via mobile outside-broadcast (OB) units. Traditionally, these involve lorry loads of equipment and attendant staff. They are slow to reach the scene of action, and not very mobile when they arrive. As a result OB units make only occasional contributions to TV news programs. More mobile, and swifter to react, is the two-man newsfilm crew. But once shot, the film has to be physically transported back to the studio and developed, before it can be shown—a very slow process.

ENG looks like the solution. It has the speed of an OB unit, and the mobility of the two-man newsfilm crew. In outward appearance, an ENG camera looks similar to a film camera. But it works electronically, in the same way as a normal studio camera. The electronic signals from the camera are recorded on a portable videocassette machine. The recordist in charge of this is also responsible for sound, which occupies its own track on the electronically recorded videotape.

As well as being recorded on a videocassette, the electronic pictures (and sound) can be transmitted instantly back to the TV station. A support vehicle can accompany the ENG crew and beam the news to the studio, either directly, or through a series of relays mounted in line of sight. Thus the pictures can be recorded at base as they are shot or, indeed, transmitted 'live' into a news bulletin.

Camera development

The capacity of two people to record electronic pictures with ENG depends upon a series of technological advances—perhaps most important of these being in the sheer weight and bulk of the equipment.

In the first 'portable' electronic cameras, an assistant had to walk behind the cameraman with some of this equipment in a box on his back—clearly an unsatisfactory arrangement.

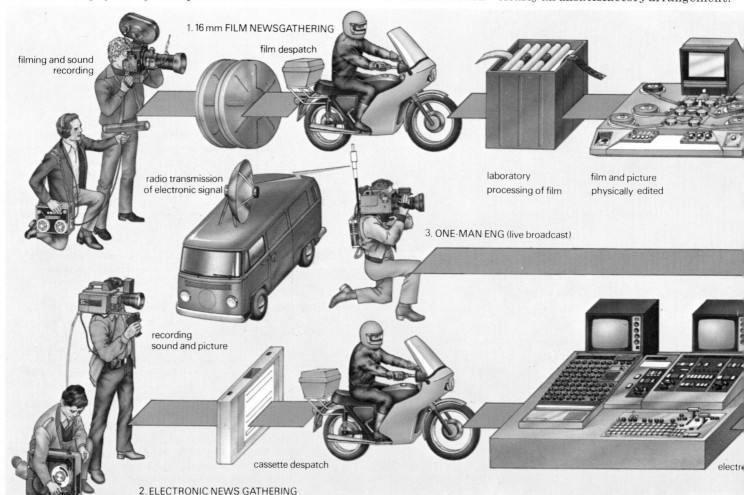

filming and sound recording

1. 16 mm FILM NEWSGATHERING

film despatch

radio transmission of electronic signal

laboratory processing of film

film and picture physically edited

3. ONE-MAN ENG (live broadcast)

recording sound and picture

cassette despatch

2. ELECTRONIC NEWS GATHERING

electr

But a modern ENG camera is a self-contained one-piece unit. All the circuitry is miniaturized, and as automated as possible. Further reductions in weight are being achieved by use of lighter casings, micro-electronic components and smaller cathode ray tubes.

Advance in the quality of the pictures that the cameras can produce is also important. Many ENG cameras have facilities for automatic aperture adjustment in response to changing light, and for minimizing the flare that arises from excessive contrasts of light and shade within a shot. They also contain all the circuitry for coding the colour picture ready for recording on cassette. ENG camera tubes tend to be more sensitive than most news film-stock, and can operate in lower light levels. (One model can shoot in the dimmest moonlight, yet compensate for any street or car lights in shot.) In general, however, electronic tubes have less tolerance to extreme contrasts (such as poorly lit faces against a bright, sunny sky) than film cameras.

With film, contrast and colour balance can be modified after shooting. For example, TV newsfilm can be adjusted when the film is converted into electronic impulses for transmission. But, traditionally, electronic pictures could not easily be altered in that way once they were recorded. The problem was that with an electronic image all the picture information was contained in one combined signal. Now, however, colour correctors have been developed which allow the electronically recorded signal to be decoded into its basic components, so the colour balance can be adjusted for transmission. This development is valuable, because ENG pictures are often taken in indifferent, fluctuating light conditions beyond the cameraman's control.

Video recorders
Equally essential for the development of ENG, was a compact, portable video recorder. The most commonly used type is a refinement of the 'U-Matic' system. This equipment was developed by Sony for industrial use in colleges and businesses and was not intended for broadcasters. But television newsmen in the US spotted the equipment's potential, and pressed the recorders into broadcasting service.

The video tape is used in 19mm (¾in) wide—wider than the 12.5mm of most home systems but narrower than the 25mm (1in) and 50mm (2in) tapes used elsewhere in broadcasting. Normal U-Matic machines can handle cassettes with up to an hour's playing time. Standard portables, or even more compact models for easier off-the-shoulder operation can accommodate 20-minute cassettes.

After the broadcasters had found a use for the equipment, Sony upgraded the system. They used the same cassettes, but accommodated more picture information, to offer higher

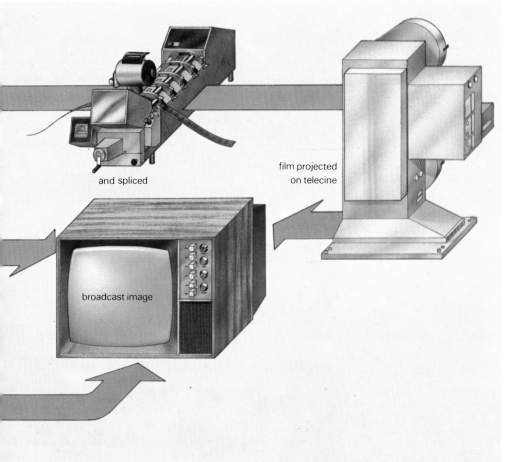

and spliced

film projected on telecine

broadcast image

A comparison of the main steps involved in film and electronic news gathering:
A news story can be captured on 16mm film (1) by a two-person unit—one to capture pictures, and the other to capture the sound. Once the film is 'in the can' it is rushed by despatch rider to a film laboratory, where it is processed. Even the quickest processing unit, using fast drying chemicals will take about 45 minutes. The film is then cut and re-ordered, to give the pictures and sequences that the program editor wants, and is stuck back together again (spliced). It then has to be projected on a telecine unit, so that the photographic images can be converted into electronic ones, ready for transmission to the viewers' TV sets.
A normal ENG crew (2) also has two people. But the camera operator also captures the sound, and the other operator records both sound and pictures on to a videocassette machine. The cassette is still carried by despatch rider, but there is no time wasting processing. Editing is still carried out (though in a different manner from film). The edited tape holds its images in electronic form, so it can be transmitted directly.
The ultimate in speed is to use a one-man crew (3) to transmit images by microwave to base, where they are broadcast live.

quality. Though these cassettes are the most common format for ENG, some broadcasters contemplate using larger machines with 25mm (1in) tape on open reels—the German manufacturer Bosch has already developed a 25mm cassette machine.

Even though picture definition is limited by the amount of information that can be crammed on to the small U-Matic tape, this problem can be resolved. There are ways of artificially 'crispening' the picture. More serious is a persistent image jitter that can be visible on such machines. Unadjusted, it could upset all the equipment throughout that chain. Fortunately, the electronics experts have devised an accessory called a *time base correct*. It smooths out all these wrinkles and brings the tapes up to broadcast quality.

Except when pictures are to be inserted 'live' into a news bulletin, program controllers need the ability to *edit* the material—to select which shots and pictures they want to include and which they want to discard. Videotape cannot be edited like film by physically splitting it. Instead, it is edited by re-recording the originally recorded shots, in the desired order, on to a new master tape. A microprocessor-based system controls the operations of the two recorders (the machine playing back the original 'master' tape and the second machine re-recording it). The controller keeps the two machines locked in synchronization, so that a particular edit can be rehearsed and repeated indefinitely. The editor can experiment by adding various new frames on to the last frame of the preceding section on the master tape.

A system called *time code* helps the editing process: it uses a separate track on the tape to record a number which uniquely identifies each frame of picture. This edge numbering system allows an individual frame to be easily found.

In the future

Over the next few years, as ENG equipment spreads even more widely among the world's broadcasters, the technique and its accompanying technology can be expected to undergo further development. For example, there will be continuing advances in the way ENG pictures are rushed back to base. Already, in the US, the microwave allocations for TV stations are becoming heavily congested. One solution, pioneered by a few US stations, is to transmit their signal across town in the form of infra-red light beams.

Many US stations have already begun to make regular use of helicopters to act as a microwave relay station to extend the range over which pictures are transmitted back to base. Sophisticated equipment is required to track the helicopter across the sky, and receive its signals clearly.

Existing camera equipment is already compact enough for solo operation (a separate recording engineer is not necessary) and it is being used in this way by some local US stations. But there will be further development in camera design. In future models the electron tube image sensor will be replaced with a 'charged-coupled device' (CCD). The CCD is a light-sensitive silicon chip, little more than a centimetre square. This new system of recording visual images means that the only limiting factor on a camera's size will soon be the lens. For some years, matchbox-sized monochrome solid-state cameras have been available for security and defence applications. When such technology can be upgraded to yield broadcast-quality pictures, ENG will be first to benefit.

Technological innovation guarantees ENG a great and developing future. But ultimately the benefits for the viewer —and the viewer's understanding of the world—depend upon the skill and judgment with which TV journalists and technicians use these new tools to gather up-to-date news.

Left: This very compact, self contained ENG unit can be carried by just one person. A UHF link transmits the pictures directly to base.

Below: In contrast to the one-man ENG unit, this more traditional OB set up requires a multitude of staff and much equipment.

TELEVISION INFORMATION SERVICES

For years, science fiction writers have been predicting that people would be able to run their lives at the push of a few electronic buttons—in particular, that all manner of information would be available on television-like screens, fitted in everyone's living room. At last, this information revolution is starting to happen.

For hundreds of years, information has been available to people by means of the printed word and picture. Books store great amounts of information in a permanent form; magazines and newspapers do much the same but are more up-to-date. But there seems no limit to the amount of information that people want to know or how up-to-date they want it. Finding out what you want to know from books can simply take too long, and even newspapers quickly become out-of-date. Television and radio help with the problem of immediacy, but you can do very little to select the information you want, and even less to alter the speed at which it is presented—you have to accept what the broadcaster is transmitting, and assimilate the information at the speed he presents it. What is needed is some way of combining the best of both systems.

It is British researchers who have done this, with two different types of system—*teletext* and *viewdata*. Many other countries are now developing similar systems. Both systems can display printed information and pictures (though normally only crude ones) on a television-like screen in an ordinary house. As with books, viewers can look at a screenful of information for as long as they like, but the information they see is held on computers, so selection is easy and rapid, and the material can be updated instantly.

The teletext system

Teletext, as its name suggests, is broadcast along with the normal television signal. It was developed almost simultaneously by the BBC (British Broadcasting Corporation) and the IBA (Independent Broadcasting Authority) during the early 1970s, out of research aimed at providing a subtitling system for the deaf which would not be visible on ordinary TV sets. The system uses the fact that the television picture and sound signal are not broadcast continuously—many times a second, there is a gap in transmission (during the blanking time: see page 12 for more details of this). The IBA was already making use of this momentary gap in transmission to add information about the source of a transmission for use by engineers at relay stations, but it was soon realized that the gaps could carry much larger amounts of information. The BBC and IBA agreed on a standard teletext format in 1974, so that reception of both services would be possible on the same TV set.

After a series of trials, the BBC's *Ceefax* ('see facts') became fully available in 1976, with Independent Television's *Oracle* (Optional Reception of Announcements by Coded Line Electronics) following in 1978. The BBC has since added *Orbit* to its second TV channel, giving Britain three fully operational television-transmitted computer information services and a lead of not less than two years over the other Western countries experimenting with the idea, including the United States, West Germany, France and Italy.

Transmitting teletext

The teletext images originate with small editorial teams, who regularly feed information into a computer, where it is formed into 'pages'. Each page is a screenful (sometimes two or three screens full) of text and 'graphics' (simplified pictures) giving, perhaps, stock market results, the latest weather or up-to-date sports results. Each broadcasting channel has its own set number of pages which are thus electronically coded and transmitted, one after the other, over and over again. Each teletext page consists of 24 lines of type, with 40 characters per line. Seven colours are available—white, red, blue, green, magenta, light blue and yellow. Each character is transmitted as a digital code signifying which combination of the 70 dots in each character display 'box' on the screen should be illuminated. A separate code, itself invisible on the screen, indicates the colour of each box.

Although the gap in the television transmission can hold quite large amounts of information, it is not long enough to allow even one full page to be transmitted in one go. Instead, several gaps are needed and so a complete page can take a quarter of a second to transmit. The two BBC services show between 100 and 150 pages each, so that a complete sequence

Far right: A page from Oracle, a teletext system, showing the weather in the form of text and simple pictures. Right: This page of information on Prestel (a viewdata system) shows clearly its two advantages. It is interactive—the computer is asking the user to tell it which group of countries he wants information on. And the pages are laid out in a tree form way—each time the user selects a page from those offered he gets more and more precise information.

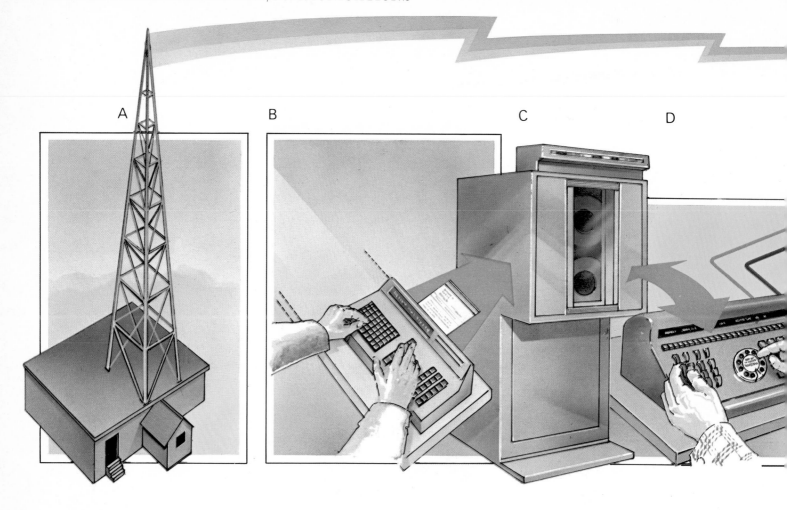

A B C D

takes about 30 seconds to transmit; the IBA service has up to 350 pages and takes about two minutes.

The first thing that is necessary before you can receive teletext pages is a special TV with circuits to decode the information in the picture gaps. As the pages are transmitted at quarter-second intervals, without stopping, it is clearly necessary for the decoder to have some way of capturing a page and leaving it displayed on the TV for as long as you want—solid state computer-type memories (see page 19) are used for this. Along with the normal TV controls are buttons to tell the decoder what service and what page (they are numbered) you want to view. The decoder then waits while the pages are cycled and, when the selected one is transmitted, stores it in its memory so that it can be displayed on the TV until you decide to have another page, or to return to conventional broadcast programs. It is the necessary wait for the page you want (the *access time*) that limits the number of pages a teletext system can sensibly hold. For example, if you call up a page on Oracle that has just been transmitted, there could be a wait of more than two minutes before it comes round again and is displayed. To reduce this delay, Oracle broadcasts the most sought-after pages—the index to the system, TV program schedules, news headlines and weather—many times each cycle, but this, of course, reduces the number of different pieces of information the system can hold.

The Teletext system is still capable of dramatic advances.

One scheme, known as *telesoftware*, would be broadcast along with Oracle on Independent Television. In effect, new Oracle pages would be added that amount to computer programs rather than pages of information. With the addition of a home computer the user could call up computer programs which would enable him to play sophisticated TV games, or use the TV as a specialized calculator, capable of working out mortgage payments, electricity consumption and other household sums. It is expected that the system could also play a role in education, providing schools with cheap and simple computer facilities.

One important point to remember about teletext is that it is a broadcast system—so you need a full television set (albeit a modified one) to receive the pages, and the system works only when television programs (including a test card) are being transmitted.

The viewdata system

By contrast, the alternative technology—viewdata—is *not* broadcast. Instead, the information pages come into the home along the telephone wires. A display screen is still necessary to view the information, but there is no real reason why this should be a full television set—though because probably all intending users of the system already have a television, it makes sense to adapt the system to make use of it.

Freeing the information system from broadcast distri-

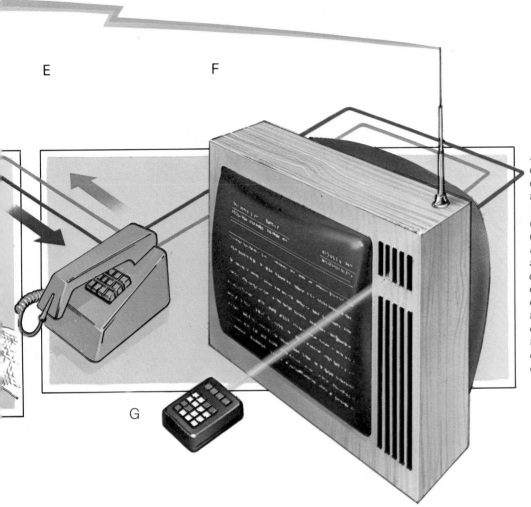

E   F

G

Both teletext and viewdata systems end up as screens full of information displayed on a special TV set (though a complete TV set is not essential in the case of viewdata). With teletext, data is stored in a computer (not shown) and is sent via a transmitter (A) direct to the TV's aerial (F), by hitching a ride in the normal broadcast signal. With teletext, data from businesses and organizations (B) is fed into a central computer (C). Space on the computer and an index to the information is provided by the telecommunications company (D). Signals travel along telephone lines (E). In both cases, the choice of information can be controlled by a keypad (G), no bigger than a calculator.

bution makes viewdata much more flexible than teletext. The most obvious freedom is that the service is not constrained to broadcasting hours—it can be available whenever the main storage computer is running, and this can be 24 hours a day. Less obvious, but much more important, is that the system can be made *interactive*—no longer does the user have to wait until the page of information he wants is broadcast. Instead, he can talk directly to the storage computer, telling him exactly what page he wants; in return the computer can provide him with it immediately. Consequently, because there is no problem of long access times, a viewdata system can hold literally millions of pages of information.

Indexing

The ability for the computer and the user to hold two-way conversations is exploited by Prestel—the main viewdata system in Britain run by British Telecom—in a special way. One of the disadvantages of screen displays is that they cannot show very large amounts of information at a time—very much less than on the page of a book, say. So much of the information on the Prestel computer is not organized in a straightforward way, with each page continuing on from

the last. Instead, a page will have a 'lump' of information on it, which will enable the user to decide which 'lump' of information he next needs to consult. The page displayed will allow the user to *branch* out to a number of other pages, one of which will contain the lump of information wanted; this is the page the user selects. The process can go on, ensuring that at each step the user has exactly the right amount of information for his needs. This progressive indexing system is called a *tree form data base*.

Viewdata generally costs the user much more than teletext. In Britain, for example, Ceefax, Orbit and Oracle are provided free, but for viewdata, the user has to pay British Telecom for the use of the computer and the phone call, and a sum to the provider of each page of information that he looks at—information providers will base their charges on the cost of providing the information, and its value to the user.

Viewdata and teletext are different, but complementary systems. Both are necessary in Britain, because few people receive their television programs via cable distribution. In countries where cable TV is much more widespread, it is possible that a single, hybrid, system could be used, combining the best features of both systems.

INDUSTRY

CALCULATORS

During the last ten years, modern technology has entirely transformed our means of manipulating numbers. Today's electronic calculators are a remarkable feat of miniaturization and performance. And they are now available at outstandingly low cost.

In operation, calculators rely on the use of a very simple mathematical process. They can only add and subtract using a simple code of one's and zero's in a system known as binary notations—see page 19. There are ways of carrying out most mathematical computations through long routines of adding and subtracting all of which would be impracticably tedious for human use, but are all in a moment's work for today's high-speed electronic circuitry.

By binary additions, or a process of subtraction known as complementary addition, the calculator carries out even advanced mathematical conversions, through routines known as algorithms. These are the particular mathematical processes it has to use in order to solve complex problems by addition alone. For example, a calculator does not remember sets of answers, as a human does with, say, multiplication tables. If asked to multiply eight by nine for example a human will remember that the answer is 72, but a calculator has to add eight nines together to produce the same answer. For a human, this process would be intolerably slow but a calculator can achieve the task in milliseconds (thousandths

of a second), or less. And it will always perform its calculations with the same degree of accuracy. More difficult mathematical processes build upon the four basic ones of addition, subtraction, multiplication and division.

However, it was not until the 1960s—when new semiconductor technologies allowed hundreds of thousands of transistors to be squeezed into small blocks of silicon, in a process known as Large Scale Integration—that the task of building a small electronic calculating machine became feasible.

Silicon chips

The Large Scale Integrated circuit, abbreviated to LSI, is a form of silicon chip. Current types contain tens of thousands of transistors which may be designed to perform various functions, from mathematical calculations to memory storage, using binary logic. One of the measurements of a 'chip's' performance is the number of active element groups (AEGs) it has, each of which may be composed of one or many transistors arranged to fulfil a specific function. Current chips may possess well over 50,000 AEGs on less than a $\frac{1}{2}$ sq cm of silicon, and it is this density of packing, which has improved dramatically over a relatively short period of development, that is the prime factor in the rapid advance of calculator ability and the rapid decrease in its cost.

Electronic calculators are organized into various functional blocks. The input unit converts decimal keyboard entries into binary code and may hold it until the information can be fed into the Central Processing Unit (CPU),

This calculator uses a liquid crystal dot matrix alphanumeric display—it can depict the entire alphabet, and special symbols, as well as ordinary numbers. A maximum of 24 characters may be stored in its memories, and remain stored even when the calculator is switched off. This allows the calculator to store short messages, and they are recalled as a running display that moves from left to right across the screen in a continuous running loop of information.
One problem with all this complexity is that the keyboard of the calculator becomes more complicated—in this example, the alphabet shares the same keys as all the numbers and commands, which can make keying-in a sequence containing both letters and numbers a tricky operation. Calculator style computers tend to use separate keys for each letter and number—but this increases the size of the machine. The increased size allows wider displays, making it easier to see long messages.

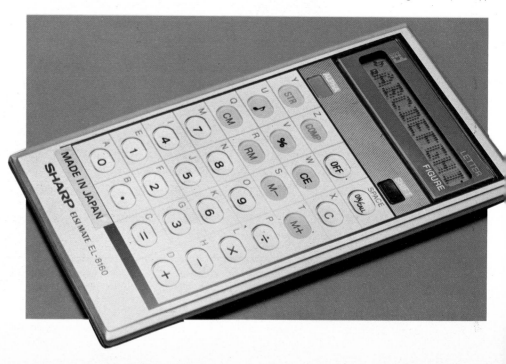

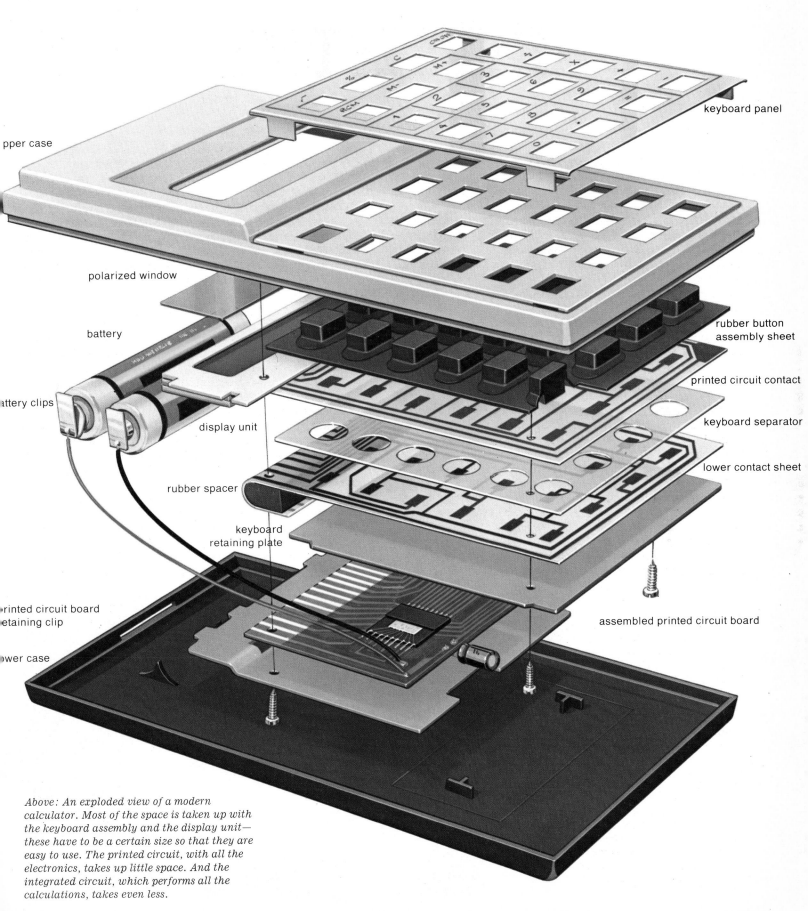

keyboard panel

pper case

polarized window

battery

attery clips

display unit

rubber spacer

keyboard
retaining plate

rubber button
assembly sheet

printed circuit contact

keyboard separator

lower contact sheet

rinted circuit board
etaining clip

ower case

assembled printed circuit board

*Above: An exploded view of a modern
calculator. Most of the space is taken up with
the keyboard assembly and the display unit—
these have to be a certain size so that they are
easy to use. The printed circuit, with all the
electronics, takes up little space. And the
integrated circuit, which performs all the
calculations, takes even less.*

which carries out all calculations. Instructional entries, such as a 'divide' command, are routed through to a read only memory (ROM see page 20) which then instructs the CPU how to process the data fed in from the keyboard. Finally, an Output Unit holds the answer from the CPU and turns it into a form that will drive a visual display that we can read.

Not all developments in calculators are connected with their mathematical computing power. Portability is a very important feature, but requires the use of small batteries of limited power; in turn this places restrictions on the display devices. Red light emitting diode (led) displays were popular for their low cost and brightness, but they used a lot of power and ran down batteries.

Attractive green vacuum fluorescent (vf) displays followed, but offered little in the way of power saving. Liquid crystal displays (lcd) have become popular for small portable calculators since they consume one hundred times less power than glowing types. The liquid crystal dot matrix display can depict alphabet characters in addition to numerals (alphanumerics), allowing calculators to deliver messages, and so on.

Calculators have already reached the point where their complexity and abilities, in mathematical calculations at least, fully meet the requirements of the market. In practice, most users require little more than basic mathematical functions and these are now available at very little cost.

Programmable calculators allow highly complex mathematical computations to be carried out repeatedly with great speed, but they must first be programmed (given instructions that will tell them how to carry out the calculations). The gap between a programmable calculator and a fully fledged computer is small and manufacturers have bridged it with small calculator-style computers that have alphanumeric keyboards, and can display one line of program or results on a liquid crystal dot matrix display.

As memory capacity increases, and displays present information in the form of words and symbols, in addition to numerals, the role of calculators is expanding. They can store messages, dates, addresses, tell the time and more—even speak and understand speech.

The calculator is becoming less of a mathematician's tool and more of a general purpose, everyday aid.

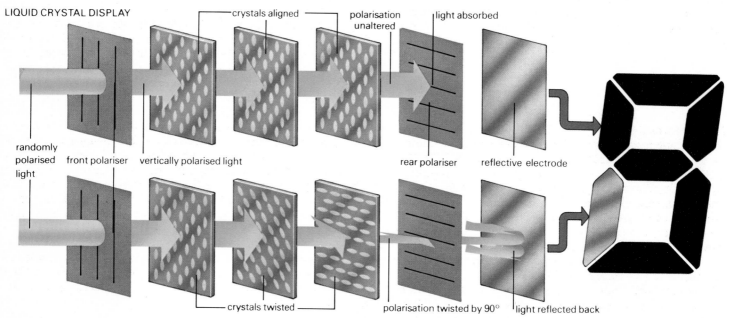

LIQUID CRYSTAL DISPLAY

crystals aligned · polarisation unaltered · light absorbed

randomly polarised light · front polariser · vertically polarised light · rear polariser · reflective electrode

crystals twisted · polarisation twisted by 90° · light reflected back

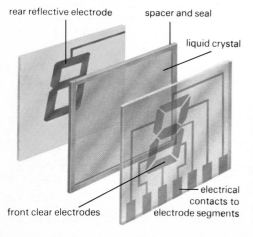

SEVEN SEGMENT DISPLAY

rear reflective electrode · spacer and seal · liquid crystal · front clear electrodes · electrical contacts to electrode segments

Left: All the numbers, and quite a few letters and special symbols as well, can be constructed by using various combinations of seven bars, arranged in a figure of eight pattern. For example, to display an eight, all bars are used; to display a zero, all bars except the centre horizontal one are used. Each of the bars has a separate electrical contact, to energize it and make it glow. Part of the electronics of the calculator are devoted to making sure the right bars of each character glow in response to the output signals of the CPU, giving the results of each calculation. This seven segment display technique is well suited to liquid crystal technology, widely used in modern calculators.

Above: An exploded view of the elements making up two segments—the top black, the bottom clear. Light falling on each segment passes first through the front polarizer which makes it vertically polarized (aligned). An energizing signal sent to liquid crystals align them so that the polarization remains vertical. The rear polarizer will pass only horizontally polarized light, so the reflective element has nothing to reflect and the segment stays black. When a segment is to be turned off, no signal is sent to the crystals, which are then orientated to twist the light so it can pass to the rear electrode. Here it is reflected, making the segment appear clear.

CRYPTOSYSTEMS

Code-makers and code-breakers have been pitting their wits against each other for thousands of years. Yet the basics of the art of cryptology have changed little in that time. One important development, though, is the advent of the computer. Computers handle and communicate masses of confidential data, which has greatly increased the work of the cryptographer. Paradoxically, the cryptographer relies increasingly on the computer to help him both make and break codes.

Cryptology covers both the making (cryptography) and the breaking (*cryptanalysis*) of ciphers. *Ciphering* is the process of changing the original message (known as the *plaintext*) letter by letter so that the enciphered message still contains the original text—but with either the position of the letters changed (*transposition cipher*) or different letters substituted (*substitution cipher*). In practice, a combination of both techniques may be employed and each process may be carried out more than once.

A transposition cipher provides, in effect, an anagram of the plaintext. A simple example would be:

```
C O M E Q
U I C K L
Y W I T H
F U N D S
T O D A Y
```

The plaintext COME QUICKLY WITH FUNDS TODAY is written horizontally and the message would be enciphered by reading downwards: CUYFT OIWUO MCIND EKTDA QLHSY. But in such a simple system, a cryptanalyst would quickly observe that reading every fifth letter would make sense of the message, and in practice more complex forms of jumbling the letters would be necessary.

Similarly, a very simple substitution cipher consists of moving each letter one place up the alphabet: A becomes B, B becomes C, etc., so that a message COME QUICKLY would be enciphered as DPNF RVJDLMZ.

This may seem more secure than a transposition, but is still easy to analyse. However, as early as the 1400s, the Italian architect Leo Battista Alberti showed that the security of substitution ciphers can be greatly increased by using different substitute alphabets for successive letters, a form known as a *polyalphabetic* cipher.

For example, instead of moving each letter of COME QUICKLY along just one place to form DPNF RVJDLMZ one might shift successive letters further through the alphabet, an additional letter at a time, so that COME might now form DQPI. In this way, any letter of the plaintext could be represented by any letter in the enciphered text. Such a cipher would require a more complex mathematical analysis to reveal its concealed message.

However, the example given would prove fairly easy for an experienced analyst to break, once he had guessed that the key was to shift each letter by regularly increasing amounts. A more secure form of this type of cipher uses a running key in the form of a random series of numbers, as in the example shown. The numbers are allocated in turn to the plaintext, and determine how far along the alphabet each letter is shifted during enciphering. For example, the first letter *T* has to be shifted by three to become *W*, but the next letter is shifted by a wholly unrelated amount (six) and so on. Similarly, the first *E* is shifted by 26, but it is impossible to calculate by how much the second *E* will be shifted: it depends entirely on what number in the key happens to occur at that point.

Ciphering machine

This type of cipher is impossible to break, provided the running key is truly random and of unlimited length. In practice, many polyalphabetic ciphers use keys of limited, even if very extended, length. Ciphering machines using long, but not wholly random, keys have been available since the 1920s —one famous example was the rotor-type *Enigma* machines used during World War II.

Most rotor machines used to have a number of interchangeable rotors, electrically wired so as to form a series of polyalphabetic substitution ciphers in such a way that any particular sequence of substitutions would re-occur only at very widely spaced intervals. The authorized recipient, with a similar machine, had only to know which permutation of rotors to use, their initial settings and (in some systems) further permutations made possible by manually adjustable plugs and sockets.

Those who devised rotor machines in the days before computers were available had every reason to believe that they would be secure against all then-known techniques of crypt-

*A polyalphabetic form of substitution cipher. The running key, which is a series of random numbers which does not repeat, shows how far forward in the alphabet each letter has to be shifted to provide the cipher text. For example, a shift of four would change the letter **A** into **D**, or the letter **X** into **B**.*

PLAINTEXT	T	H	E	Q	U	I	C	K	B	R	O	W	N	F	O	X	J	U	M	P	S	O	V	E	R
RUNNING KEY	3	6	26	1	12	8	3	11	16	1	19	8	18	25	6	14	17	15	2	5	7	18	1	13	14
CIPHER TEXT	W	N	E	R	G	Q	F	V	R	S	H	E	F	E	U	L	A	J	O	U	Z	G	W	R	F

analysis. In the event, the German and Japanese Enigma-type machines were broken during World War II and methods of attacking the similar American Hagelin machines have been disclosed. This is the penalty for not using keys that are truly random.

In practice, cryptosystems are not always required or expected to be unconditionally secure—provided that they either delay sufficiently the recovery of the plaintext or involve the code-breakers in an unduly massive and unjustifiably costly operation. A system that meets such objectives is often called *computationally secure*.

Digital techniques

During the past decade or so, a major change has taken place in cryptography involving the adoption of high-speed digital techniques directly related to those used in computers.

In a sense, all cipher transmission by radio or land-line involves the use of digital codes. Morse or teleprinter codes, for example, are digital codes. But until recently, messages were enciphered as letters *before* being transmitted digitally.

Modern practice is to convert the plaintext directly into digital form. The next step is to encipher the stream of digital 'bits' by combining them with a digital running key.

With high-speed digital electronics, the whole process can be carried out in 'real time' at virtually any required speed, both text and speech, and transmitted without delay. In effect, the use of a digital key creates a form of polyalphabetic substitution cipher. If the running key is truly random and of unlimited length, such a system would be unconditionally secure. In practice, a pseudo-random key is derived from a relatively short (typically under 100 bits) key, but so arranged that a long key is produced before any repetition of the sequence occurs.

The way in which the pseudo-random sequence is produced has been shown to have a highly important effect on the security of the cryptosystem. It is now known that many of the original digital systems, although resulting in enormous numbers of enciphering possibilities, can be broken readily by relatively small computers—provided that some part of the message (for example the opening address) is already known to the cryptanalyst.

So there are now cryptosystems in which each block of text and its key is subjected to a very complex series of changes, including transposition of order and substitutions based on long key sequences derived from the original few. Such systems are generally believed to be computationally secure against all but the most determined attack.

In 1977, the American National Bureau of Standards en-dorsed the Data Encryption Standard (or DES) algorithm (the program which tells the computer how an encoding key, once generated, is to be used) based on a 56-bit key. Each data block passes through 18 data-manipulation stages in which 16 different enciphering keys are derived from a 56-bit main key to provide a total key length of 10^{17} bits. Users of this system use a *published* algorithm but have paired key-generators for which the main 56-bit key is kept secret. This key-generator can take the form of a small, sealed module, but it remains important to keep this secure. DES cryptosystems are now being adopted by commercial and banking organizations, and nobody has proved publicly that they can be broken. But some cryptographers believe that the 56-bit main key is too short for the system to be regarded as secure against a really determined and massive computer attack, an alternative type of cryptosystem—*public* key cryptography—has been advocated as potentially providing greater security. In addition, the system reduces problems of key distribution and management.

In these systems, the functions of enciphering and deciphering are separated. The sender does not hold the deciphering key but enciphers the message for a specified addressee, using a published list of enciphering keys. It is claimed that such cryptosystems should be capable of providing computationally secure systems without the problem of having to distribute (and keep secure) key generators.

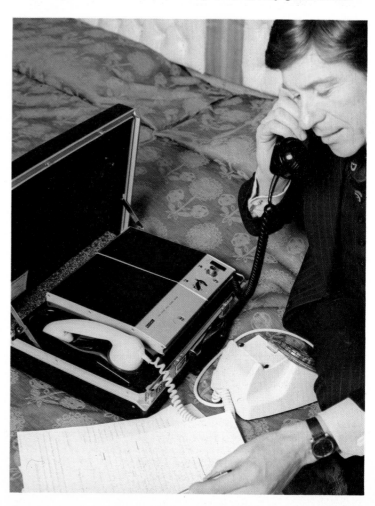

Right: Digital electronic techniques have revolutionized the encryption of messages. It is now no longer necessary for messages to be written down, coded laboriously, and sent by telegraph or Morse to a recipient who then has to decode the message before it can be read. There are now audio encryption units which can change speech into digital form, then encode and transmit the message as fast as it can be spoken. A similar machine at the recipient's end will decode the message into audible speech just as fast—this is real time encryption. Shown is a device that employs frequency dispersion techniques —a sophisticated form of 'scrambling' that permits a total possibility of 4.2×10^{10} combinations.

Above: Boris Hagelin, inventor of a six-disc rotor machine that produces 101,405,850 different encoded letters before it repeats itself. Sales of the machine to the US military made him a millionaire.

Right: The Data Encryptation Standard of the American Bureau of Standards. The plaintext is divided into 4-bit digital types, and subjected to an initial permutation. The key generator then applies 16 different keys in turn to the initial permutation. (The diagram shows a sample of three keys.) At each stage, the digital form message (shown here symbolically in letter form) is further transposed. A final permutation gives a total of 18 stages of substitution and transposition.

US DATA ENCRYPTION STANDARD (DES) SYSTEM

64-bit key

56-bit derived key

derived generator key

key no 1

key no 2

key nos 3-15

key no 16

64-bit plain text

initial permutation

manipulation stage 1

manipulation stage 2

manipulation stages 3-15

manipulation stage 16

final permutation

TOP SECRET

64 bit encrypted text

HOLOGRAPHY

Since photography was first invented, people have tried to produce a three-dimensional photograph. The nearest thing using conventional photographic techniques is the Stereoscope—but this is very limited in its three-dimensional effect. A method of producing a true 3-D picture was conceived and predicted by the British scientist Dennis Gabor in 1947. But the practical demonstration of the technique had to await the invention of the laser in the early 1960s.

A conventional photograph yields only a flat, two dimensional representation of the subject. The end product of holography, the *hologram*, captures the three dimensional information about the scene—that is, it includes information on depth. Using holography, a reconstructed scene can be viewed from a range of angles—you can 'look round' the side of the photograph.

Laser light

Light sources used in conventional photography, for example sunlight or electric lighting, emit radiation over a wide frequency range (ultra-violet to infra-red)—white light contains a jumble of different frequencies (colours). Because of the disordered nature of white light it is not possible to record information about the depth of a scene. To record depth information the light source must be of a single frequency, that is, *monochromatic*; and with each wave in phase with every other wave—coherent. Such a beam of light is emitted by laser devices—see pages 36 and 37. Details of the depth of a scene illuminated by a laser beam are contained in the phase relationships of the waves arriving at the holographic recording plate. A wave arriving from a more distant part of the scene will 'lag' behind waves from closer points

and it is this information about lagging and leading waves which is recorded in a hologram.

To record this information the laser beam is split into two parts: one is directed at the scene from which the reflected beam (called the *object beam*) is derived and the other part is aimed directly at the recording plate (this is the *reference beam*). Where the object and reference beams meet at the plate they will interact or *interfere*.

When the crests of two waves coincide, an enhancement of light intensity or amplitude occurs through the wave energies adding together. This process is called *constructive interference*. When the crest of one wave coincides with a minimum position, or trough, of a second wave a reduced intensity is obtained—*destructive interference*.

The recorded pattern of the beam at the holographic plate contains both amplitude and phase information about the object beam. A conventional photograph records only the amplitudes of the light arriving at the film.

The developed holographic film, or hologram, looks nothing like the recorded scene. If the object being recorded is a simple flat reflecting surface the resulting interference pattern shows a series of light and dark bands, whereas the pattern produced by light reflected from a single point on an object consists of a series of concentric rings. A hologram of a complete object or scene is a highly complex pattern of superimposed circles from the many points on the object that reflect light.

Normally, the hologram is developed as a transparency. To construct an image of the original scene this transparency must be illuminated with a beam of coherent light similar to that used as the reference beam in recording. Viewed from the other side, a complete reconstruction of the original object is obtained.

In detail, the reconstructing laser beam is modulated (changed) in amplitude and phase as it passes through the

Holographic images are a common feature of laser light shows—three dimensional images can be projected, hanging in mid air.

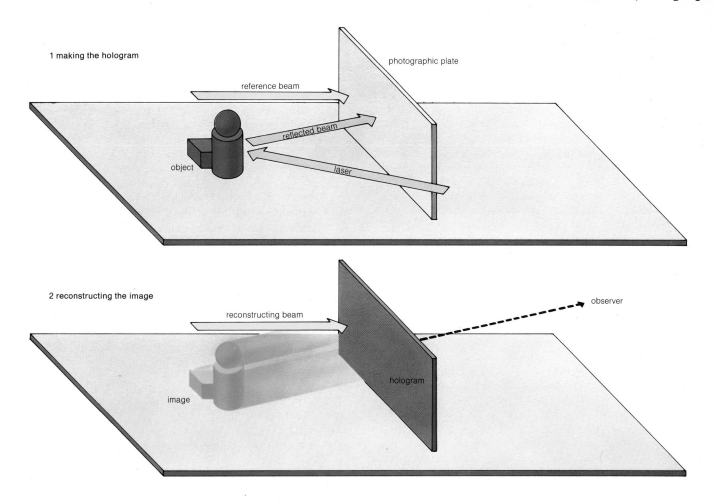

1 making the hologram

photographic plate

reference beam

reflected beam

object

laser

2 reconstructing the image

observer

reconstructing beam

hologram

image

hologram transparency in such a way that it resembles the original object beam. This produces a *virtual image* of the object which, to the observer, appears to be behind the hologram. It is this image which can be seen by the eye.

A hologram produced using a single laser beam will give only a monochrome (single colour) reconstruction of the object. It is possible, however, by choosing three laser beams giving three different frequency beams (corresponding to the three colours—red, green and blue) to record and reconstruct a scene in full.

Applications

Three-dimensional pictures look exciting the first time they are viewed, but the technology of producing them is not particularly far advanced. Holographic movies have been made, but the image quality is low, and the films can be seen by only a few people at a time.

Holography, however, is not limited to the making of 3-D photographs, and there are many other practical uses for the technique which are proving more important. One industrial use exploits further the phenomenon of interference, to reveal extremely small differences between the dimensions of a master object and a manufactured copy. The reflected object beam from the copy is directed at the hologram of the master. When viewed from the other side, interference patterns (fringes) will exist whenever there are differences between master and copy. Each fringe from a given reference

A hologram records complete three dimensional information about an object, using a laser beam which is split into two parts. One part is aimed at the object, and the reflected light from this which hits the photographic plate (hologram) is called the object beam. The other part (the reference beam) is aimed directly at the plate. To view the image, a reconstructing beam, identical to the reference beam from exactly the same direction, is needed.

point indicates a size difference of one half wavelength between the test object and the master. Use of a typical laser light source enables size differences of 0.0003mm to be readily detected.

This technique is also used to detect the minute changes in the dimension of an object when it is subject to mechanical stress—car manufacturers, for example, use it to detect the stress imposed by tightening cylinder head nuts, and there are many other possibilities.

Because a highly directional reference beam is needed to make a hologram, it is possible to store many images on just one holographic plate—simply change the direction of the reference beam between exposures. This certainly gives the possibility to store many pictures on one plate, but there is a much more important use—that of storing large amounts of computer-type information in a small area. But like many other advanced computer memories (page 19) much research still needs to be carried out.

LASERS

Laser 'death rays' that can blast satellites and planes from the skies infallibly are an awesome idea, but the range of potential laser applications is so spectacular as to make even the death ray seem rather mundane. From steel drills to surgery, from holograms to hi-fi, the development of the laser is only just beginning.

'A solution looking for a problem', was how the laser was described in its early days and those words have proved prophetic. Every day a new use is discovered for laser technology and lasers are now applied to processes which had always seemed adequate until the laser highlighted their limitations. The laser is illuminating not only the frontiers of scientific, nuclear and space research but also proving an invaluable tool in normal everyday activities. Even when they are used simply for display purposes lasers can be impressive. As the lasers are switched on, the night sky is crossed by shafts of light, so pure in colour and so straight and well-defined that they seem like tangible rods.

All these properties stem from the process of stimulated emission which gives the laser its name—Light Amplification by Stimulated Emission of Radiation. Although the first laser was not built until 1960, the concept of stimulated emission had been proved theoretically by Albert Einstein as long ago as 1917.

Stimulated emission

Light is energy in the form of electromagnetic radiation, emitted by atoms. To despatch a photon of light (a packet of radiation) the atom must be *excited* by an input of energy that boosts its energy level in a series of definite steps called *transitions*. The wavelength of the photon of light emitted as the atom falls back toward the *ground state* (the unexcited state) depends upon the transition the atom goes through. Transitions at high energy levels produce long wave infrared light while those at low levels produce short wave ultraviolet light; visible light comes in between.

In a conventional light source, such as a bulb filament, the emission of light is a rather haphazard affair, as atoms fall

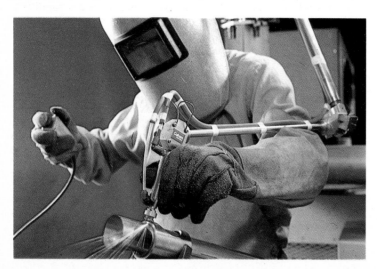

back *spontaneously* towards the ground state radiating photons with a tremendous variety of wavelengths in all directions and at random intervals of time. However, Einstein showed that an atom can be stimulated to emit a photon before it would do so spontaneously if it is hit by a photon whose wavelength corresponds exactly to the transition the atom will go through.

When this happens, the atom will emit a photon of precisely the same wavelength as the photon that hit it and the two will move off travelling in the same direction at the same time (that is, *in phase*). Since this process can be repeated on many other atoms, producing an identical photon each time, each of which can contribute to further stimulations, *stimulated emissions* can thus amplify one photon into a whole stream of photons of exactly the same wavelength, travelling in phase and in the same direction—this is known as *coherent* light.

When Theodore Maiman produced the first working laser in spring 1960, he used a solid rod of ruby. It has since been shown that almost anything, from krypton to carbon dioxide can be made to operate as a laser.

With the ruby rod laser, the 'lasing' material is not the ruby itself but the chromium impurities it contains. The idea is to flood the rod with light energy from a flashtube to ensure that the majority of chromium atoms are in an excited state—in fact, at the third transition level. The atoms soon lose energy and decay to the second level and, for a while, there are more atoms in level two than level one (the ground state). Sooner or later, one of the atoms in level two decays spontaneously to the ground state, emitting a photon which can stimulate other transitions to the ground state, as Einstein predicted.

Of course, this process of stimulation begins not just with a single atom, but with many thousands and the light, although of the same wavelength, shoots off in all directions. However, light that is not travelling parallel to the laser axis, or close to it, is lost through the walls of the crystal before it can stimulate much emission. On the other hand, light that is parallel to the axis is trapped within the rod. Mirrors at each end reflect the light back and forth, and as it traverses the length of the rod it stimulates more and more emissions, all moving in the same direction. By making one of the two end mirrors less than perfectly reflecting, the amplified light can escape from the rod in a thin, parallel coherent beam of great intensity and a single wavelength. Different lasing materials from ruby will produce light of different wavelengths.

The coherence of laser light and its parallel beam endow it with great value as a measuring tool. If the laser is aimed at a target its distance can be calculated very accurately from the reflected light through a variety of techniques such as measuring the time the pulse takes to return.

Armies have not been slow to realize the potential of such system for accurate range finding. Laser measuring has been adopted in civil fields as well, and the variety of people who benefit from its accuracy increases all the time. Engineers, for instance, use lasers to check the alignment of aircraft wings or to dig tunnels straight.

A flexible laser beam guide for complex cutting.

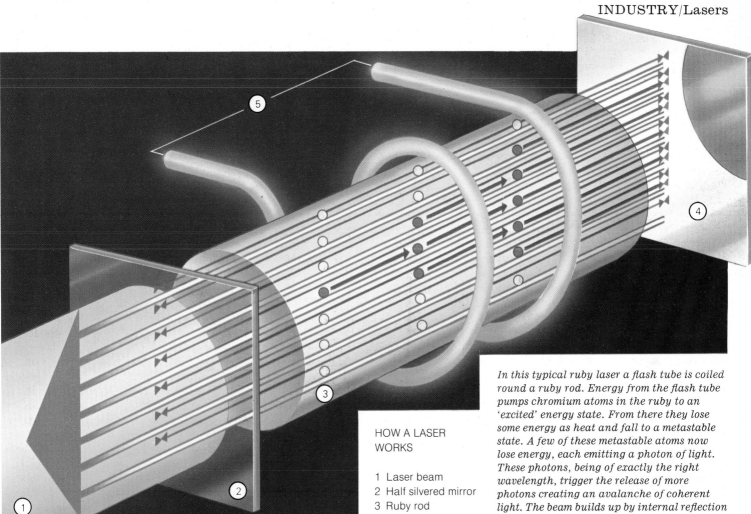

HOW A LASER
WORKS

1 Laser beam
2 Half silvered mirror
3 Ruby rod
4 Silvered mirror
5 Flash tube

In this typical ruby laser a flash tube is coiled round a ruby rod. Energy from the flash tube pumps chromium atoms in the ruby to an 'excited' energy state. From there they lose some energy as heat and fall to a metastable state. A few of these metastable atoms now lose energy, each emitting a photon of light. These photons, being of exactly the right wavelength, trigger the release of more photons creating an avalanche of coherent light. The beam builds up by internal reflection at each end of the ruby rod before emerging at the half-silvered end.

With lasers, geodesists can plot the location of continents and land masses to within a few metres—a hundred times as good as with conventional techniques. It will soon be possible to detect whether continents are indeed drifting relative to one another at a rate of a few centimetres every year.

Another valuable feature of lasers is that they can concentrate energy to perform remarkably delicate and small-scale operations. It is possible now to focus the beam of a laser down to a diameter of 0.025mm. The microelectronics business in particular benefits from this facility, though it is valuable in many other areas.

But perhaps the most obviously beneficial use of micro-welding is in eye surgery. It is fairly common, for instance, for the retina of the eye to become partly detached through disease or injury, which can cause blindness. By choosing a particular wavelength, laser light can be made to pass through the front of the eye without any effect and focus on a small spot on the retina. The concentration of energy will then produce scar tissue to 'weld' the retina back into place firmly and with the minimum of damage to the retina.

The potential for lasers in medicine seems to grow every year with the testing of lasers for making bloodless surgical incisions, for the destruction of tumours, for the eradication of skin blemishes and for the removal of tooth decay.

While the laser has a great future for mending bodies, it also seems to have a potential for breaking them. Ever since its introduction, military establishments have been watching the progress of the laser closely and examining the feasibility of a death ray. At first it was thought that lasers would be too weak to be effective over anything but very short range. Indeed, an Austrian scientist provided theoretical proof that death rays were never really likely. However, the development of powerful carbon dioxide lasers changed the picture. These lasers could soon punch a hole through virtually every substance in a matter of seconds at close range and could burn the clothes off a soldier at quite a distance.

The short wavelength (that is, high frequency) of a laser makes it ideal for communication. The relatively low frequencies used by radio waves can carry only limited amounts of information: lasers make it possible to use the higher frequencies of light to carry much more information. The problem is that, unlike radio waves which can be broadcast in all directions and be picked up out of sight of the transmitter, lasers travel in a single straight beam. They are also badly affected (like all light) by atmospheric conditions, so they are unlikely to be used for general broadcast. But for rapid communication between two points their value is escalating each year. Many telephone companies are beginning to replace conventional cables with fibres carrying laser light which carry far more calls of much higher quality.

NUCLEAR REACTORS

The process of nuclear fission—'splitting the atom'—is one of the staggering achievements in scientists' understanding of the very fundamentals of how matter is created and how it can be manipulated. The nuclear reactor of an 'atomic' power station is one of the ways in which the forces released during fission can be controlled for people's good.

The basic building blocks of all substances are atoms; there are a hundred or so types of atom, and all substances are made up from various combinations of these. An atom consists of a relatively heavy central nucleus carrying a positive electrical charge, and negatively charged electrons whirling in orbit around it. The major components of the nucleus are the *protons*—positively charged particles—and (except in the case of ordinary hydrogen) neutrons which carry no electric charge. The protons give an atom its identity—each type of atom has a different number of protons. However, the number of neutrons in any particular type of atom may vary. For example, uranium exists as several variants or *isotopes*. Each isotope has 92 protons, but different isotopes have different numbers of neutrons. The two main isotopes are uranium-238 with 146 neutrons, and uranium-235 with 143—three fewer (the number after the word 'uranium' tells you the number of neutrons and protons altogether in that isotope; the same scheme is followed for describing isotopes of other elements).

Splitting an atom
Nuclear fission occurs when the nucleus of an atom splits into two parts—some nuclei can be made to split by bombarding them with 'spare' neutrons. For example, uranium-235 can be made to split into xenon, which has 53 protons and 89 neutrons, and strontium-90 (38 protons and 52 neutrons). This leaves, among other things, two extra 'spare' neutrons. Even so, the total mass of all the products of the fission process is not quite as much as the original mass. The difference appears as heat energy, according to Einstein's famous equation $E = mc^2$, where E is the amount of energy produced, m is the lost mass and c is the velocity of light. It is this heat energy that nuclear reactors harness.

Because the velocity of light is such a large number, even a tiny amount of lost mass will produce a huge amount of energy—this is what makes nuclear energy so important. But there are many problems in harnessing it safely and efficiently.

Chain reaction
One difficulty is in keeping up a steady fission reaction. If one of the spare neutrons released when uranium-235 undergoes fission ruptures another uranium-235 nucleus, and a neutron from this nucleus ruptures another, and so on, then a *chain reaction* is said to have been achieved. If, following each fission, more than one other nucleus is ruptured, then the reaction gets out of control—which is what happens inside an atomic bomb. On the other hand, if after a fission no other nucleus is ruptured then the reaction will stop. For generating controllable energy, therefore, a chain-reaction

that just maintains itself at a critical state is required.

Natural uranium contains only about 0.7 per cent of uranium-235—far too little to produce a steady reaction. In a nuclear reactor, this problem is solved in two ways. First, the proportion of highly fissile uranium-235 can be increased by a process called *enrichment* until it is high enough to ensure criticality. In practice, two or three per cent uranium-235 is adequate. Second, the free neutrons produced by fission can be made more efficient at rupturing other fissile nuclei. The most effective neutrons are those that have lost much of their energy; they are called slow or thermal neutrons. So the neutrons in a reactor are deliberately slowed down by using a *moderator*, such as graphite, which has the same effect on the speed of a neutron as water has on that of a bullet.

As well as encouraging the reaction, there must be some means of controlling it and, when the need arises, stopping it. This is done with control rods, which pass straight through the reactor core. The rods are made of a material that readily absorbs neutrons. By varying the number or length of control rods in the core, the chain reaction can be managed as required. Most reactor safety systems work on the principle of 'dropping' all the control rods in an emergency in order to shut down the reactor.

Radioactivity
One of the popular worries about nuclear reactors is the threat of radiation. Unfortunately, the products produced during fission are mostly highly radioactive, giving off alpha, beta and gamma radiation—all of which, in sufficiently large doses, can be fatal. Some of these nuclear waste products can remain dangerous for centuries. And as

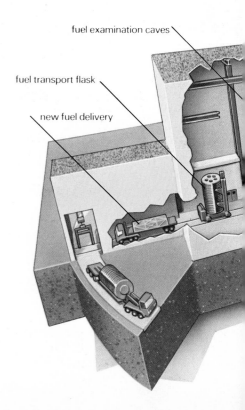

fuel examination caves

fuel transport flask

new fuel delivery

The Fast Breeder Reactor is most efficient at producing energy. Unlike other types of fission reactor, it actually produces more fuel than it burns. But this means that it also produces much highly dangerous radioactive nuclear waste, the disposal of which remains a serious problem.

yet no real solution to the problem of safe disposal has been found—which means that part of the cost of nuclear power is the legacy of danger we leave our children. The greatest amount of effort has to be put into the design and operation of a nuclear power station to make it safe—but some opponents of nuclear energy say that stations can never be made safe enough.

Other factors that affect a nuclear reactor design are the cooling of the reactor core—which would otherwise reach a temperature similar to that of the Sun—and the conversion of the energy produced by nuclear fission into electricity. In practice, these two factors are often closely linked. Although controlled nuclear reactions are the product of a recent technology, nuclear power stations are little more than glorified steam engines. The cooling material, be it liquid or gas, is simply run through a heat exchanger where water is boiled, producing steam which is used to turn turbine generators.

Magnox reactors

The first nuclear power station to be linked with a national grid was the Calder Hall Reactor in Britain. Switched in during 1956, this 50MW reactor is of the type generally known as *Magnox* reactors—named from the material used to clad the fuel elements. The fuel used is natural uranium metal and the moderator is graphite. Cooling is by means of pressurized carbon dioxide gas, which passes through a heat exchanger. In order to contain the pressurized gas coolant, the whole core is contained within a vast, sealed pressure vessel. In the earlier designs, this was either a welded steel cylinder or a sphere but, to meet demands for a larger and more powerful reactor, welded steel was replaced by pre-stressed concrete which also helped protect operators better from high-energy gamma radiation.

Refuelling Magnox reactors is a complex process because their cooling systems are pressurized. Another disadvantage is that natural uranium deteriorates too rapidly in a reactor

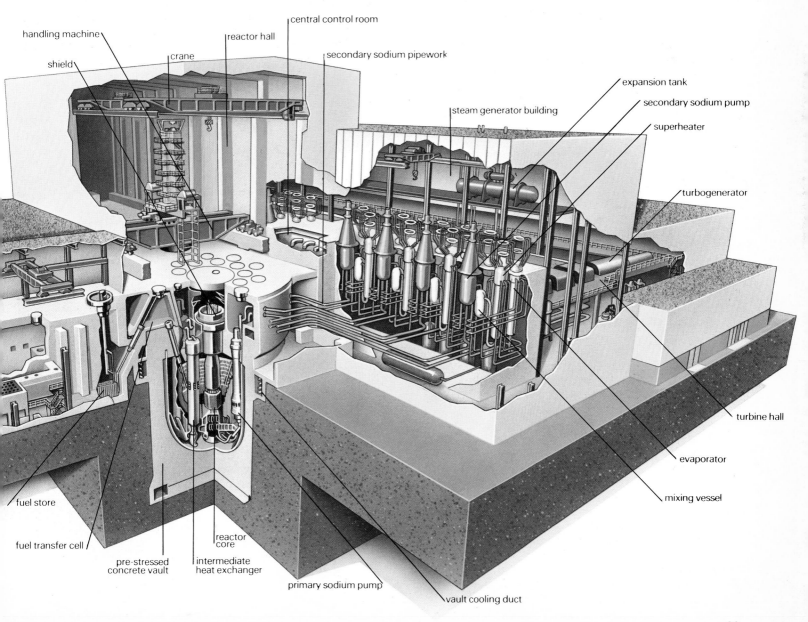

39

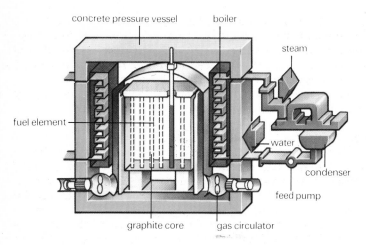

Above: In an Advanced Gas-cooled Reactor, the moderator is graphite, and the coolant is pressurized carbon dioxide gas contained in a concrete pressure *vessel which also acts as a shield. The generating turbine is powered by steam raised in a boiler, in much the same way as in a conventional power station.*

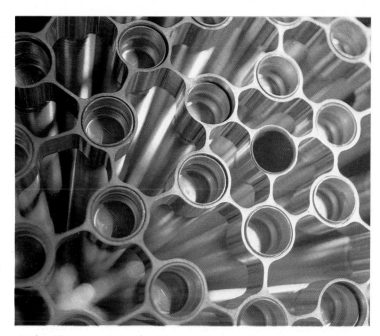

Above: The fuel arrangement in an AGR. The reactor is fuelled with tiny slugs of enriched uranium oxide enclosed in pins clad in stainless steel. In order to *increase the surface area of the fuel and improve heat transfer, the pins are only one centimetre (less than half an inch) in diameter.*

core to be useful. Finally, cooling is a problem—the Magnox cladding material is relatively susceptible to heat, and carbon dioxide is in any case not a very good coolant—and this limits the amount of power a Magnox reactor can safely produce.

Other types of reactor

Some of the failings of the Magnox reactor have been overcome in the *Advanced Gas-cooled Reactor* (AGR). This reactor employs an oxide of uranium which lasts longer. Stainless steel cladding is usually used. This can cope with higher temperatures, but is a bad reflector of stray neutrons, so the uranium oxide has to be more enriched.

Considerably more advanced is the *High Temperature Gas-cooled Reactor* (HTGR). The HTGR contains no metal in its core—instead, ceramic materials are used. The fuel (very heavily enriched uranium oxide and thorium, a material that becomes fissile after nuclear bombardment) is made in the form of tiny spheres which are then coated with carbon and silicon and embedded in the moderator (conventional graphite). This arrangement gives both a much more even mixture of fuel and moderator, and a very compact reactor core. Four times smaller than a Magnox reactor, it is equally powerful and its fuel lasts 30 times as long.

Because extremely high temperatures can be achieved within a HTGR, it is possible to pass the cooling gas (helium) directly through a gas turbine and eliminate the steam cycle altogether.

Fuelled by slightly enriched uranium in zircaloy (an alloy of zirconium) fuel pins much like those used in AGRs, the United States' *Pressurized Water Reactor* (PWR) employs ordinary water as the coolant, moderator and reflector. To cope with the heat it is pressurized to approximately 150 atmospheres. The fuel has to be enriched because, unlike graphite, water readily absorbs stray neutrons. Because the coolant is also the moderator, and thus has a direct bearing on the state of the nuclear reaction, PWRs always have emergency cooling systems to cope with main system failure.

The overall efficiency of the PWR is not very high. Its major drawback is that by relying on pressurized water as the coolant, reactor temperatures must be kept reasonably low.

Just as the AGR followed on the heels of the Magnox reactors, the *Boiling Water Reactor* followed the PWR. In this reactor the water performs all the functions that it does in the PWR, but by being allowed to boil and produce steam it also directly drives the generators. Such a system, however, has its losses as well as its gains: although boiling water is much more efficient at heat transfer, the resultant steam is a less efficient coolant. This means that large amounts of steam must not build up in the system.

Two other reactors, *the Canadian Deuterium/Uranium* (CANDU), and the *Steam Generating Heavy Water Reactor* (SGHWR) use heavy water as the moderator. Heavy water is deuterium oxide, a compound of oxygen with *heavy hydrogen* (an isotope of hydrogen which has one proton plus one neutron in its nucleus).

Unlike other reactors, the Liquid Metal *Fast Breeder Reactor* (FBR) utilizes fast neutrons because, although less efficient than slow neutrons, when they do succeed in causing a fissile nucleus to rupture, more neutrons are generally released. To overcome the basic inefficiency, 'fertile' material is added to the fuel—elements like uranium-238, which while not fissile themselves can be converted to fissile material following collision with a fast neutron. In doing so, a point will be reached where more fissile material is being generated than is being lost—they actually produce more fuel than they consume. FBRs are extremely effective in producing energy, but they are difficult to operate safely.

OIL MINING

The main focus of attention in oil production over the last few years, particularly in the UK, has been in oil from the sea bed. But looking for oil on land is just as important, and with new techniques aimed at unlocking hitherto untapped resources, oil mines could be part of the answer to the energy crisis.

Certainly a large part of the reason for the rapid rise in the importance of oil mines (in which the oil 'locked up' in oil shales and tar sands is extracted) is that the price of oil is now so high that it justifies the great cost of extracting it by mining techniques. And the amount of oil available in this way is vast.

Estimates, of course, vary widely but the total amount of 'conventional' oil left in the world is put at about one and a half million million barrels—and this is not expected to last very long. The oil shales of America alone are estimated to contain at least two million million barrels—and worldwide, the quantities are expected to be about ten times as high. Oil mines will not, of course, make oil energy cheap and abundant, but they will put off the evil day when the oil runs dry altogether quite significantly.

Oil shales and tar sands contain their precious fluid not in pockets of 'pure' oil but distributed like water in a sponge.

Sucking the oil out of this 'sponge' is made more difficult because it is very sticky—up to 50,000 times less fluid than conventional crude oil. The usual way to make the oil flow is to heat it to several hundred degrees first. The synthetic crude, or *syncrude* produced from this oil can then be used to manufacture petrochemical feedstocks for products ranging from polyesters to detergents.

Fort McMurray

In Canada, mining projects are underway to exploit the oil locked up in gritty tar sands stretching under huge tracts of northern Alberta. The four tar sands in Alberta are estimated to contain a total of more than 900 thousand million barrels of heavy oil. A barrel is 35 gallons (42 US).

Canada's first modern oil-mining plant came into production in 1967,—about 40km (25 miles) north of Fort McMurray in Alberta. The tar sands at Fort McMurray are located about 15m (50ft) below the surface. They occur as a 30m thick layer and comprise 11 percent bitumen—a tar derived naturally from petroleum. The tar sands are exposed by digging long, narrow trenches to remove overlying earth and rock— a conventional strip-mining technique. The sand is then piled up along the trench so that huge excavators can transfer it to conveyor belts that, in turn, carry it into a processing plant.

Once inside the Fort McMurray plant, the crude bitumen is separated from the sand using a hot-water process devel-

Esso's 'huff and puff' project at Cold Lake, Canada, relies on superheated steam in order to make the thick, deeply buried oil flow. Once at the surface, the oil is refined into a light synthetic crude oil.

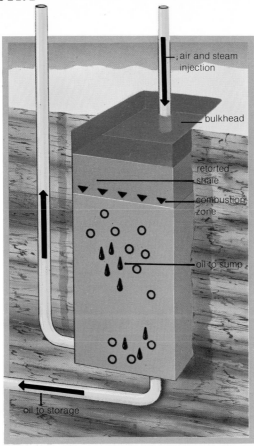

Far right: The shale oil plant at Fort McMurray—a huge capital investment, with no guarantee of a high return.

Right: Mining shale oil by retorting the shale underground reduces the problems of handling the large quantities of materials mined by conventional methods, as well as reducing the cost. Retorts are constructed in groups, the shale heated and then the resulting oil is piped off.

air and steam injection

bulkhead

retorted shale

combustion zone

oil to sump

oil to storage

oped by the Alberta Research Council. The tar sand is exposed to superheated steam, which causes an oil slush to drop down into a water bath below. The sand and gravel sink, but the oil remains afloat and is skimmed off the surface. For every 60,000 barrels of bitumen collected (a day's production) 55,000 barrels of syncrude can be upgraded, ready for refining.

A Canadian-US company—Kruyer Tar Sands Development —has successfully tested a new process that extracts 99 percent of the oil from the sand using an oleophilic (oil-loving) material. The process is claimed to be more economical and safer environmentally because it does not pollute the process water. Tar sand mixed with water is moved over a conveyor-belt screen made of the oleophilic material. The oil sticks to the conveyor; sand and water fall through.

At Athabasca—the site of Alberta's largest tar-sand deposits—strip-mining techniques enable access to only about five percent of the oil. The other 95 percent lies much too deep at up to 760m (2500ft) below the surface. Like the other Albertan deposits—around Cold Lake, Peace River and Wabasca—the Athabasca tar sands require other techniques. Essentially, the oil is heated underground and then made to flow to the surface.

Huff and puff

The most promising technique—called 'huff and puff'—is already producing about 200,000 barrels of oil a year in an experimental process at Cold Lake. Huff and puff involves injecting high-pressure steam, sometimes mixed with hot water, into an oil well over a period that can last for weeks. The well is then shut in and the reservoir around it allowed

to soak. After perhaps several weeks more the well is reopened to produce oil that is sufficiently fluid for natural pressures to force it to the surface along with the water pumped down earlier. When the output begins to decline, the whole process is repeated. Far from being straightforward, the process is technologically challenging. It is estimated that to produce 160,000 barrels of oil a day, 8000 oil wells will have to be drilled at Cold Lake, giving an average 100 wells in every 1.6 square km (one in four acres).

Almost 600,000 barrels of water a day will have to be pumped down the wells over periods of about five days and it will be necessary to re-inject steam every time the fluid temperature during production drops below 65 degrees C (149 degrees F).

In the USA, research into the production of shale oil on a large scale dates back to the late 1940s, using strip-mining techniques to prepare oil for heat treatment, or underground fires or steam injection to raise the oil. During the late 1960s, however, the main focus shifted to applying a combination of both methods. Occidental Petroleum is developing a project to produce 50,000 barrels of oil a day at two mines in Colorado.

The modified in situ process, developed by Occidental, involves the complete removal of about a quarter of the shale to create a series of individual underground caverns, some as large as a 30-storey office block. Conventional explosives are used to shatter the shale left inside the caverns. Then the rubble is burnt where it lies in its own separate retort. The oil vapour produced subsequently condenses and drains off through exit shafts leading to storage tanks.

ROBOTS

Robots will revolutionize the production lines of the world during the next few years. Strong, accurate and reliable, they can handle dangerous or monotonous work at a constant high standard, releasing human workers from hot, noisy and dirty environments. And they can do sophisticated tasks too—already they are used for assembling products as varied and complex as cars and wrist watches.

Robots, of course, are still machines, still dependent on their human instructors. But the amount they can be taught is increasing dramatically and so is the control they exercise on their work. In loading and unloading a metal forging press, for example, a robot has to pick up a white-hot billet from the chute and place it in the press. Then, after the forging, it has to remove and place the billet in a trimming press for the next operation. A signal tells the robot that a hot billet is available at the pick-up point. The robot takes the billet and moves to the front of the press, then stops. Then it signals the press 'are you open?'

If the answer is 'no', the robot waits at that point in the program. If the answer is 'yes', the robot places the part in the press. Now it says to the press 'don't close, I've got my hand in'—and the press is prevented from operating. Once the part has been placed in the die and the robot arm has withdrawn, the robot tells the press that the forging cycle can be started. A similar routine is followed to remove the part from the forging press and load it into the trimming press.

Compared with the robots of science fiction, perhaps, such machines seem elementary. But as micro-technology develops, especially in the electronics and computing fields, rapid advances must be anticipated. Communications between robots and other machines in the work place (known as 'interlocking') is done by means of micro-switches and other sensors. This kind of apparatus is becoming more ingenious so rapidly that it seems to change every month.

Robot anatomy

The first industrial robot made its debut in industry in the United States in the early 1960s, and the design of robots themselves has not changed greatly over the two decades. The industrial robot basically comprises three parts: a hydraulic power unit, a mechanical unit and a control system. The hydraulic power supply may be integrated into the body of the mechanical unit or be a separate, free-standing unit. It provides the energy to operate the various axes or 'arms' of the robot. The hydraulic piston may drive directly, through gears, levers or other linkages, or a hydraulic motor may be used if direct rotary motion is required.

The 'machine' part of the robot, the mechanical unit, provides the means of picking up and manipulating components or manoeuvring various tools such as spray guns or arc welding torches. A typical mechanical unit can operate along three main axes. On the horizontal plane the 'arm' of the machine both extends and retracts; a swing motion, like the waist of a man, permits the arm to revolve about the base; and the tilt axis, the 'shoulder' of the robot, allows the arm

to be raised and lowered over a very wide range of angles.

At the end of the horizontal arm are three further axes, which provide the same motions as the human wrist. A gripper attached to the wrist works like the human hand and can take many forms. A simple two finger unit can open and shut to pick up simple parts, while vacuum cups can pick up sheet components such as glass. Special tools such as spot-welding guns or other devices can also be attached to the wrist.

One critical component in the robot system is the servo valve. This is electrically controlled: its job is to control and direct the flow of hydraulic fluid around the robots' circuits, and so make the robot move.

The 'brain' of the industrial robot, the control unit, is usually housed in a separate cabinet and connected to the mechanical unit by electrical cables. Earlier machines made use of magnetic tapes to store information but today's industrial robots employ solid state silicon electronics and

Above: This Swedish robot is programmed to move through 200 set positions in the course of polishing stainless steel sinks.

Left: At Volvo's assembly shop, a robot applies glue to car body components.

increasingly make use of micro-computer technology. This has eliminated the mechanics of tape drives, and consequently improved the reliability of robots.

Teaching a robot

The control system is made up of three main parts: an amplifier to drive the servo valve; a logic or instruction element to control the robot's functions; a memory element to store program information; and feedback elements which provide information about the position of the various robot axes.

The operation of any robot can be reduced to two fundamental modes, known as *point to point* and *continuous path*. In the first, the path by which the robot travels from one position to another is unimportant; the quickest route is the best. The control unit is required only to record the points to which the robot must move; the exact path and speed are determined by the robot's own operating limits. An example would be loading a machine. For continuous path operation, however, the route is all-important—for example, when the robot must follow the contours of an object, as in paint-spraying. Here, the control unit is required to record in detail the path of the robot hand and also its speed.

If the robot is going to load a machine, it must first be taught how to do it by a human operator. The operator drives

the robot's arm by means of a 'teach' control to the point where the part is to be picked up. When he has carefully positioned the hand at this point, he operates a record button and the 'address' of that point is recorded in the control system memory. He moves through the entire program in this manner, leading the robot hand to each desired point and pressing the record button each time to record the

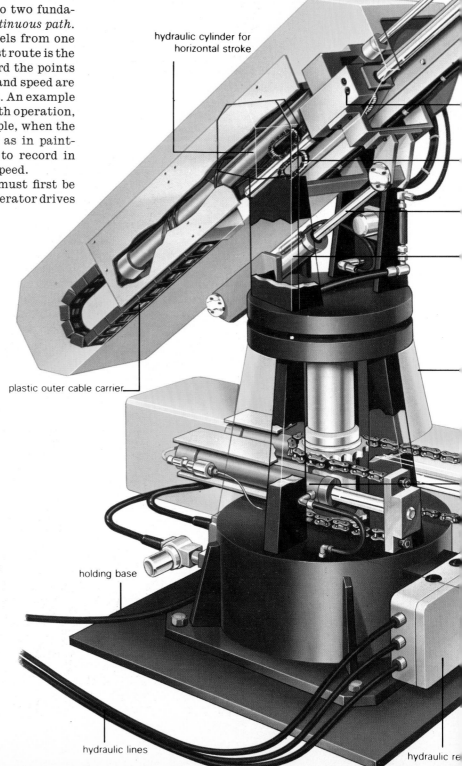

hydraulic cylinder for
horizontal stroke

plastic outer cable carrier

holding base

hydraulic lines

hydraulic re

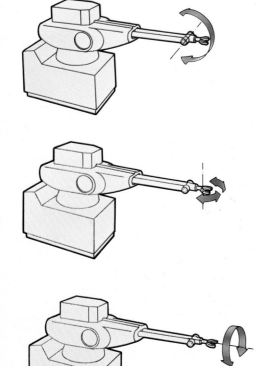

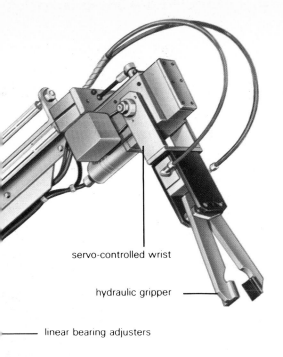

servo-controlled wrist

hydraulic gripper

linear bearing adjusters

linear bearings

hydraulic cylinder

tilt cylinder

outer casing

swing cylinder

Left: The large picture shows the component parts of a typical, multi-purpose robot. This sort of machine could be trained to perform most of the functions that a human could carry out with one arm—and in many environments where a human could not go.

Far left: The small drawings show the axes of movements that the wrist of a one-armed robot is capable of. From top to bottom, the wrist can swing vertically, horizontally or can rotate. A variety of grippers can be attached to the wrist—for example, vacuum cups or 'finger' units.

Below: The three axes of movements of the arm (again, comparable to the movements in a human). From left to right, the arm can swing horizontally, vertically or it can move back and forth.

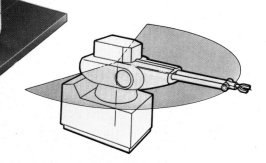
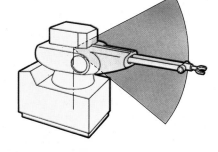
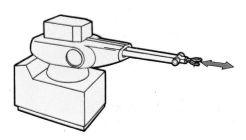

address. In playback the robot arm will move to each of these points in sequence.

For an application such as paint spraying, it is necessary to record the operator's movements continuously as he moves the spray gun. The operator 'teaches' the robot the task by actually moving the mechanical unit arm, with the spray gun attached to the wrist, to perform the task. All his movements, including the trigger actions as he turns the spray gun on and off, are faithfully recorded in the control system memory.

The feedback element

All this 'teaching' information is transmitted to the control system by a 'feedback element'. This is linked to the mechanical motion of the particular axis by gears, belts or some other means, and sends the control system an electrical signal proportional to the position of each axis. Thus, when the robot is stationary at one address, that information is passed to the control system and the address of the next point in the program is sent back down the line. The servo amplifier responds to this signal by driving the spool of each servo valve in the appropriate direction to cause the correct motion of each robot axis.

Even as the axes move, the feedback elements are also moving and sending back to the control system positional information. This is processed in the logic element and as the address of the next point is approached, the command signal diminishes and the servo valves progressively close, slowing down the movement of the axis. When the address received from the feedback element in each axis exactly matches the address for that point as stored in memory, the command signal is reduced to zero, the servo valve is centralized, hydraulic fluid ceases to flow and the axis motion is stopped.

What of the future? The mechanics will probably not change greatly, since they are mainly designed to fit a broad range of tasks, but the robot will become much cleverer; the micro-processor will see to that. The ability to process more and more information at faster and faster speeds will give the robot an increasing level of intelligence. Vision will enable the robot to 'look' at a part and to recognize, for example, which way up it is or to discriminate between good and bad components. Touch sensors fitted to the gripper fingers will allow the robot to vary the force with which it holds each object so that delicate objects can be handled safely. Improved software and programming techniques will give robots more precisely defined 'personalities'—and voice recognition techniques will make them react to such rudimentary but vital commands as 'STOP!'

SECURITY SYSTEMS

The sight of security guards at a modern factory or industrial complex is ever more common. With their uniforms, and perhaps weapons, you might be forgiven for thinking that their only role was to prevent unauthorized intruders and thefts. But a company can suffer loss and damage through many other causes, such as fire or the failure of important machinery. Today's guards look after all aspects of security, and may have a centralized computer-controlled system to help them.

In one of the most modern systems, piece-meal approaches to security problems have been replaced by a single system, providing total security. At the heart of it is a computer, controlled by a single operator. Acting on information from a range of sensors, the computer monitors smoke, fire, flood, intruders—or an employee making a legitimate entry. More than 3000 individual sensors, distributed throughout a building or plant, can be linked to the computer and separately monitored through just one cable. And the computer is not fooled if a wire to a sensor is cut: that action itself would raise an alarm.

Many different types of sensor are available. One of the most common types is a *magnetic reed detector*. Current passes through two fine metal reeds, housed in a glass tube, fitted in a small plastic case to the frame of a door or window. Another case containing a magnet is fitted next to the reeds, but on the door or window itself. As long as the two are close together, the magnet on the door holds the reeds together and the current flows. When the door is opened more than a small distance, the magnet moves away from the reeds, causing them to spring apart and break the circuit.

Perhaps the simplest alarm sensor is a *microswitch*—a tiny electrical switch actuated by a plunger. The plunger is held down by, for example, a closed door. If the door is opened, the plunger springs up and breaks or closes a circuit.

Acoustic detectors are small microphones that register sound. They can be made insensitive to certain sound frequencies but sensitive to a particular sound—breaking glass, for example. *Photoelectric detectors* respond to light; *temperature detectors* respond to heat or cold. In some alarms, temperature and vibration detectors are incorporated in a device called a *safe limpet*, which can detect someone trying to cut open a safe. Temperature detectors can also be used to guard against fire, or to sound an alarm if a refrigerator fails.

The most advanced sensors are those that detect movement. *Ultrasonic detectors* send out high frequency sound waves, which are reflected in a predictable pattern by furniture and walls. The sensors can detect if the reflected sound pattern is disrupted, as it would be by the movement of an intruder. *Microwave detectors* work in the same way, except

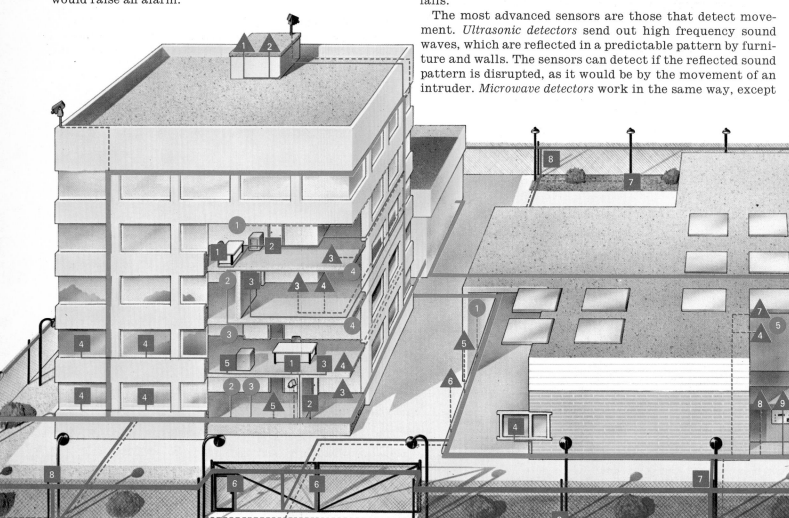

that they use radio waves. Microwave and ultrasonic sensors can be used to scan a large area, such as the floor of a factory, or a large open-plan office. They are simple in operation and difficult to defeat, especially because they are easily concealed.

Infra-red sensors make use of a narrow beam of low-frequency light, just invisible to the naked eye. The beam can be projected over long distances to give security to, for example, a suite of offices or factory perimeter. The beam is transmitted to a target, which reflects it back to a receiver. If the beam is broken, the alarm sounds.

Clusters of sensors are wired to control boxes along a single cable. The computer interrogates each control box more than once a second by sending a coded electrical pulse down the cable. Each control box has its own code, which can be identified among a string of codes reaching the computer. All codes can be interrogated simultaneously. The process is known as *multiplexing*. Activation of any of the sensors is immediately registered by the computer, which displays an alarm signal on one of the visual display units (VDUs) in the control room. Some computers automatically display a map showing the location of the sensor that signalled the alarm.

The programming of the computer determines what happens next. In the most sophisticated programs, the computer can take a host of actions: it can automatically switch on sprinkler systems to douse a fire; switch on television moni-

An integrated security system allows the operator to receive in seconds correct and precise information from a variety of sources which may be thousands of feet apart. The input to the control centre can be from any form of sensor—human, electronic, fire, movement or temperature; there are virtually no limits. Extensive use of multiplexing techniques means that the signals from the various inputs can be sent in code from down the same cable. Over 3000 individual sensors can be connected and separately monitored by one two-wire cable, no thicker than a pencil. The illustration shows examples of the types of security equipment that might be needed to protect a large complex. The system is activated by various sensors arranged on branches off the central multiplexing line (shown in red). The system is designed to automatically send for assistance from the police or fire authorities so that help arrives with the minimum of delay. However, when possible, the operator activates remote controls to enable him to deal with the problem without leaving his desk. This means that in an emergency the operator has the opportunity to continue monitoring the rest of the complex, and at the same time take suitable protective action.

AN INTEGRATED SECURITY SYSTEM

PURPLE
1 Tank Water Level
2 Lift Monitor
   Master Switch—
3 Lighting
4 Heating
5 Energy
   Management
6 Environment
   Monitor
7 Pollution Monitor
8 Heat Monitor
9 Process Monitor
10 Output counter
11 Flow Indicator
12 Electricity Supply
   Monitor

GREEN
1 Fire alarm bell
2 Manual break
   glass
3 Inert gas release
4 Infra Red Smoke
   Detector Beam
5 Sprinkler system
6 Magnetic release
   Smoke Stop
7 Heat Detector
8 Fire Monitor
9 Smoke Detectors
10 Water Flow

BLUE
1 Personal attack
2 Protective switch
3 Movement
   Detector
4 Break Glass
   Detector
5 Safe Alarm
6 Access Control
7 Security lighting
8 Surveillance
9 Invisible Ray
10 Vibration Detector
11 Roller Shutter
   Monitor

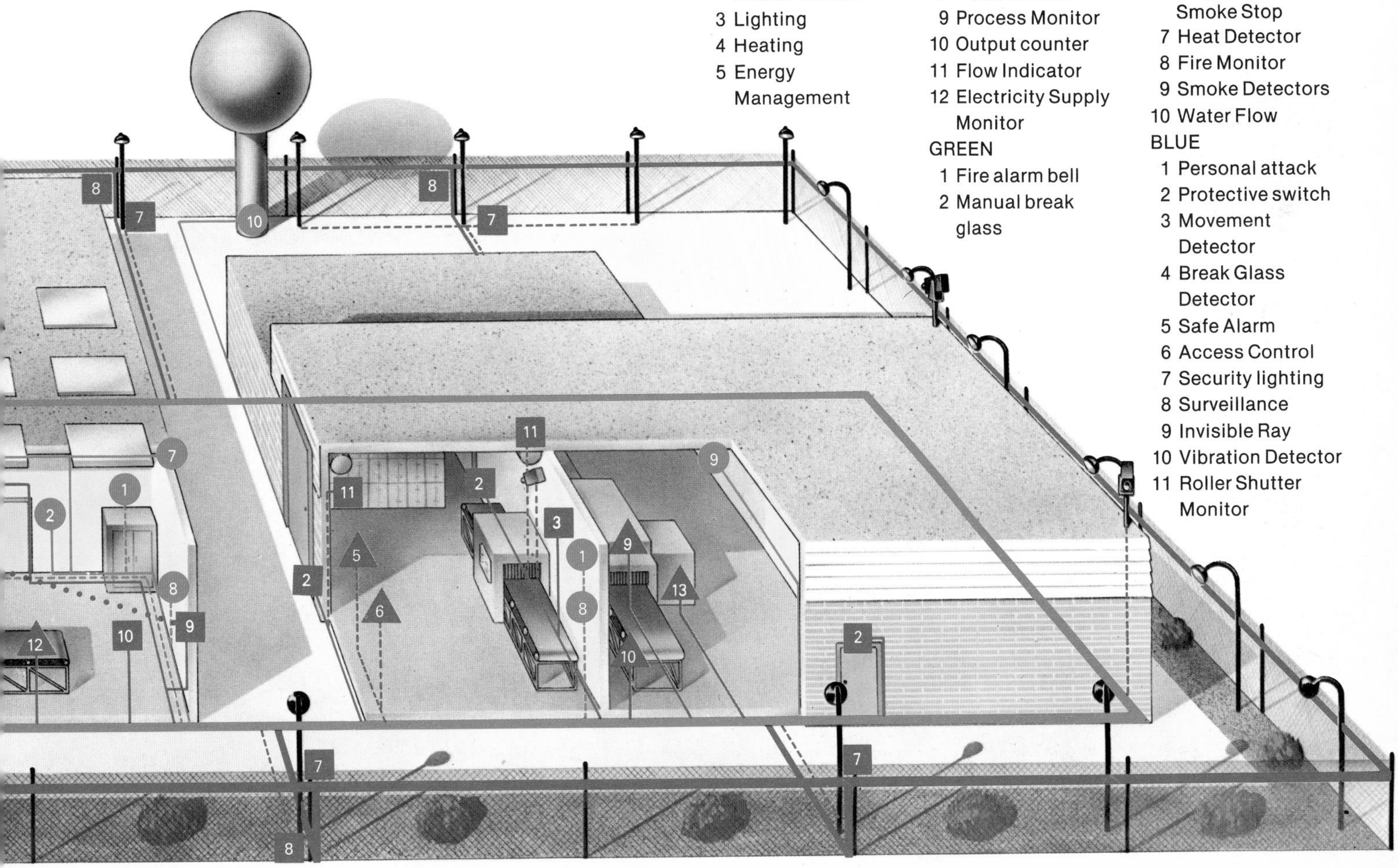

A range of sensors and control units—
Top, left to right: Ionization detector; outdoor movement detector; fixed temperature detector.
Bottom right: Card access control unit.
Bottom far right: Temperature rate-of-rise detector.

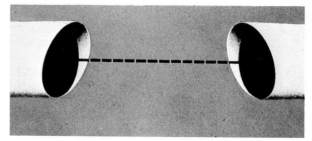

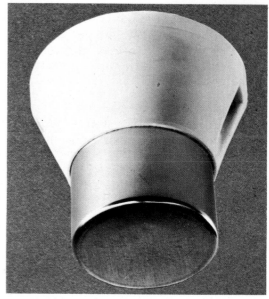

tors and operate them to show the exact location of the alarm on a VDU in the main control panel; switch on lights and control their direction so that an intruder can be seen on a television monitor; open, close or lock doors; flash instructions to the operator on the VDU so that there is no danger of human error through panic.

Not only can all these functions be available to a single establishment, such as an office-factory-warehouse complex, but one computer can control security, environment and production functions at several different locations, provided they are linked by a direct telephone line. The continuous monitoring and control of essential parameters, such as temperature, pressure, liquid level, and pollution, is just as important (in, for example, an oil refinery) as the detection of an intruder.

In industry and commerce, a great diversity of information, including confidential reports, trading and finance records, computer programs and other trade secrets, need to be protected against theft, sabotage or industrial espionage. At the same time, the building needs to be easily accessible to staff and authorized visitors. For total security, a controlled-access system can be linked to the same computer that controls intruder alarms or environmental and production processes.

The building to be protected will be partitioned into zones that require different standards of security. The zones have electrically controlled doors. Authorized personnel are given a card with a magnetic message (as on a bank cash-

point card) or a personal radio transmitter. The zone doors open only for people who have the correct card. Each door has a card reader, which opens the door to the holder of a valid card for just a few seconds.

For extra security, there can be two sets of doors around a man-trap, with floor pressure sensors that can detect an intruder who slips through the first door after it has been properly opened. Systems that use a radio transmitter to unlatch a door can be fitted with a second sensor, which sounds the alarm if anyone without a transmitter passes by. Cards that control access either have just a number, which the computer checks against its records when the card is presented, or they can be coded with detailed information.

The use of coded cards has the advantages of simplicity and cost. The same cards can be used to monitor the working hours of employees, keep records of how they move around the building, or even debit expenditure in a staff canteen! A computerized security system is the closest that technological man has come to total security. But a system has to be safe not only against human attack from within and without, but also from human error and inquisitive tampering.

In 1972, a security system that restricted access to a West German company was gradually dismantled and made useless by the employees themselves. They did not deliberately vandalize the system, but could not contain their curiosity about how it worked. They peeled apart their identity cards and pried open the new devices around them. Disillusioned, the management sold off its system to a local car park.

SOLAR POWER

The Earth receives only a tiny fraction of the sun's huge and continuous flood of energy. But this quantity is equivalent to the output of 173 million of the world's largest power stations—most of which are burning costly and irreplaceable fossil fuels. How can we plug into this vast, free source of energy?

Probably the simplest way to use sunshine, on an individual scale, is for domestic water heating, by allowing it to heat water flowing through *flat plate collectors*—which are basically radiators working in reverse. The collectors are usually placed on the roof and angled to catch as much direct sunshine as possible.

Storing solar energy

Solar power is not easily applied to space heating because the sun does not necessarily shine when it is most needed. To overcome this problem, considerable research is being directed towards finding a method of storing solar energy.

Energy can be stored in water (for example, hot water storage tanks) by raising its temperature. Other materials can be used in the same way to store heat, including the walls of a building, but to store a lot of heat energy this way requires masses of material, and very good insulation.

Another possible form of solar energy storage is to utilize a reversible photo-chemical reaction. First, solar energy is used to 'drive' a chemical reaction to produce a high-energy end product—a process akin to charging up a battery. Then, at a later stage, the high-energy product is allowed to revert to its original constituent materials. At this point, the stored solar energy is released as heat, which provides a valuable supply of space heating.

A major drawback to this system is the high cost and bulk of the starting materials. Many of the chemicals are obscure compounds based on a ring of carbon atoms. They include derivatives of benzene and napthalene and they are both costly and complex to produce. A possible photo-chemical

storage reaction is the conversion of norbonadiene (NBD) to quadricyclane. NBD could, in theory, be produced cheaply on a large scale from two cheap compounds—acetylene and cyclopentadiene. NBD has the capacity to store heat efficiently—1200 joules could be stored by each gram. But, although it absorbs light, it does not do so at the wavelengths present in sunlight. Certain additives can change the absorption pattern so that NBD is persuaded to absorb sunlight, but the chemistry here is sophisticated, and is still being researched.

Photo electric conversion

An alternative way to trap solar energy is to produce electricity directly from sunlight using a photo-electric cell. Photo-electric conversion relies upon the property of semiconductors to liberate free electrons when hit by photons (particles of light). A cell is normally made of a thin slice of silicon sandwiched between two layers of metal, such as silver. When light falls on the cell, electrons are liberated, and flow across the layers to produce a small current.

At present, the cost of solar cells is prohibitive—to power an average-sized house would cost around £50,000. In America, researchers are looking for simpler ways of producing single crystals of silicon photo-electric material. One method under test is to pull a thin crystal from molten silicon. In Britain, a 'glass cell' is being developed. By using two sheets of glass, spraying each with a mixture of chemicals, and then dipping them in a copper solution, a photo-electric glass sandwich has been produced in the laboratory.

To generate power successfully on anything other than a domestic scale, it is usually necessary first to concentrate the amount of solar energy falling over a wide area on to a small spot. Mirrors are widely used for this. A suitably shaped mirror (in a curve described mathematically as a parabola) will reflect rays of light from the sun to a single point (the focus).

An extension of the parabolic dish concept is the central receiver collector. This involves focusing sunlight by a series of flat plate mirrors, rather than by one large curved mirror. In order to keep the sun's light focused at maximum bright-

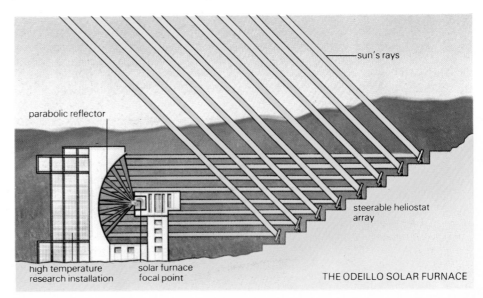

A diagram of one of the most powerful solar power stations in the world, at Odeillo in the French Pyrenees. Banks of sun-tracking mirrors reflect light onto a vast, highly-polished parabolic reflector. The reflector then focuses the energy onto a 3000 degrees C solar furnace—a temperature hot enough to melt metals.

parabolic reflector

sun's rays

steerable heliostat array

high temperature research installation

solar furnace focal point

THE ODEILLO SOLAR FURNACE

This diagram shows a solar power station similar to that at Albuquerque (see opposite page). It uses a circulating sodium heat exchange system. Hot sodium accepts heat much more readily than water, making it an ideal heat transport medium. Solar-heated sodium gives up its heat to water in the heat exchanger, and this water is raised to steam. From this point on, the generation of electric power, by using the steam to drive a turbine generator, need be no different from conventionally fuelled power stations.

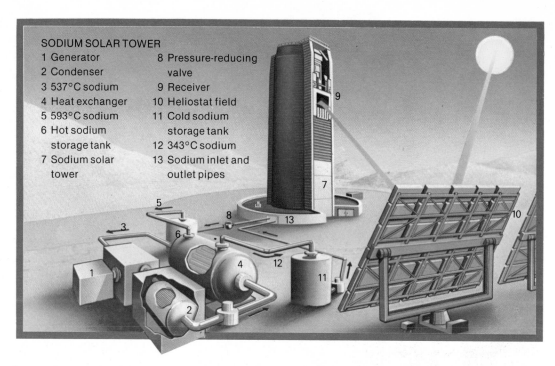

SODIUM SOLAR TOWER
1 Generator
2 Condenser
3 537°C sodium
4 Heat exchanger
5 593°C sodium
6 Hot sodium storage tank
7 Sodium solar tower
8 Pressure-reducing valve
9 Receiver
10 Heliostat field
11 Cold sodium storage tank
12 343°C sodium
13 Sodium inlet and outlet pipes

ness, thousands of mirrors may be used, each individually steered to follow the sun.

One of the biggest central receiver installations is at the Sandia Laboratories, New Mexico. The solar 'power tower' has 1775 mirrors on 300 modules. Their focus—as bright as a thousand suns—is at a steamraising boiler, perched on top of a 200-foot tower. A comparable installation has been built at Odeillo in the Pyrenees.

The $21 million New Mexican tower does not actually generate power. It is a test bed for an advanced solar-powered, 100-megawatt electric generating station. This power plant would have an 'orchard' of 20,000 mirrors. Each mirror would be individually steered to follow the sun, and focus reflected light onto a liquid sodium receiver. Sodium has the property of accepting heat much more readily than water, so only a relatively small quantity need be used. Thus, a compact sodium-carrying heat system could channel heat away from the tower to a nearby boiler room. At the boiler room, the liquid sodium would be used to heat water, raise steam and thus drive turbines.

Fluorescent collectors

Researchers are also experimenting with fluorescent collectors to concentrate solar energy before directing it onto photo-cells which could greatly cut down the number, and therefore cost, of the photo-cells needed.

The system (called a luminescent green-house collector) consists of a framework of two sheets of glass or plastic separated by a layer of fluorescent dye. The dye absorbs sunlight over a broad range of wavelengths, and then re-emits it at lower wavelengths. The trick to the system is that it is arranged so that the light re-emitted by the dye is trapped between the plates. It bounces from side to side, being internally reflected until it reaches the edges—in much the same way as light travels down an optical fibre. In the collector, one edge carries the solar cells, the other a reflecting coat-

ing. Sunlight (even if it is diffuse) is concentrated and channelled towards the cells. And there is no need to move the device to keep it pointing at the sun, since it accepts indirect sunlight.

The output from a solar power station is at the mercy of the whims of weather. But a network of solar power plants could be spread over a wide geographical area, and united by a grid —the idea being that the total energy flow will then be quite reliable, even if the output from the individual plant fluctuates quite widely. Soviet scientists at the All-Union Scientists' Research Institute of Current Sources have studied this idea. They find that such a grouping scheme could be four or five times more reliable than a single power station.

Another solution to the erratic supply of solar power is the hybrid solar and hydro-electric plant. It has been proposed, and patented, for the Horse Mesa Dam of the Salt River project in Arizona. The idea is to dovetail the supply of solar energy (available only at certain times during daylight hours) with hydro-electric power (available 24 hours a day).

To achieve this, the rate of water released from the dam is continually altered, according to the performance of the solar-powered turbine—which in turn depends on the available sunshine. When solar power is low, hydro-electric power is stepped up, and when solar power is high, hydro-electric power is stepped down.

Solar cells are now turning up in increasing numbers of specialist applications. They are particularly useful to power remote installations, far away from power lines. Perhaps the most ambitious remote solar-power installation is the Australian 580-km microwave telephone link between Alice Springs and Tennant Creek in Northern Australia— the first large installation of its kind in the world. Dialling a call courtesy of the sun may not be cheap (the system has so far cost about $100,000), but solar energy has, at least, extended communication to the Australian outback.

Above: A solar power station at Albuquerque, New Mexico, showing the huge array of steerable mirrors in the foreground, and the solar tower in the background onto which they concentrate the reflected sunlight.

Left: Solar furnace at Mont Louis in France. The gigantic size of this solar collector, and the high polish on the mirrors, is evident here.

LEISURE

ANIMATION

Until recently, the process of realizing animation from first rough sketch to completed cartoon film was a technical feat that required the equivalent of a lifetime's work. 'Animal Farm' for example, one of the first serious full-length animated feature films, required 250,000 drawings that took 300,000 hours to create. But now the computer is changing the face of animation.

The basis of any type of animation is a series of individually drawn 'still' pictures, each one varying very slightly from the last. If these pictures are shown, one after the other very quickly, then the small differences merge into one another, and the illusion of movement is created. It is the same technique, in fact, that is used to make any moving picture—a movie film consists of a huge number of still photographs (frames), each capturing a split-second of the moving action, joined together so that they can be shown at the rate of 24 frames a second.

The process of animation

To make an animated film, therefore, the usual technique is to use a standard movie camera to photograph the animation pictures one frame at a time, with the picture being changed between each frame. So just one minute's animation can require a staggering 1440 different drawings. The first stage in the laborious work is a written outline of the story and action, which sets out the main dramatic development of the film and the part the various characters will play in it.

At the same time, the directing animator makes preliminary sketches which begin to reveal the visual nature of the characters and the general graphic design and appearance that the film will assume. The colour design, as well as the graphic style, has to be tried out and decided.

Finally, the visual essentials—representing not only basic expression seen from varying viewpoints but also the scale of the various parts of the face, head and body—are agreed upon. Then the 'key' or principal animators can begin to work out the actual movement which will bring the characters to life.

The key animators, while working within the framework of their technical instructions, are primarily responsible for the creative vitality of the film. They draw the key phases of movement—the ones that determines the life and expression of a character—while their assistants complete the 'in-between' phases.

Computer aids

Computer systems are now being developed to speed up these laborious processes. Perhaps the simplest form of help is the computer-controlled animation, or rostrum, camera. This consists of a camera poised directly above a flat board, the camera table. The camera can be raised or lowered in a precise series of tiny steps; the table can also be moved back and forth, or rotated.

Normally, the cameraman follows a precise sequence of camera or table movements to achieve the effect of pans, zooms or cartoon movements. For example, a cartoon figure might be photographed in a sequence of running movements. To add realism, a background 'cel' (sheet of plastic) is moved

lead screw

camera cradle

field indicator
and track scale

camera

anti-reflective baffles

Right: An animation camera is the same as a live-action movie camera, except that it exposes each frame of film separately—in the same manner as a stills camera. Each 'cel' is placed on the animation table and individually photographed. Not all representations of movement in an animation need come from physically different cels—the camera can move closer to or further away from the table between shots, giving the impression that a character is coming nearer or moving away. The table can also be moved. When the film has been developed, it can be processed and edited—cut up and re-ordered, with only the best sequences used. Unlike live-action film, though, animations generally require little editing, because much more accurate pre-planning can be done.

counterbalance weight

platen

animation table

turntable

racking handle

camera support
columns

bevel gears

Below: Farmer Jones, a character from Orwell's 'Animal Farm', starts to run. The whole strip occupies less than one second of motion. The animator gives extra emphasis to important facial expressions and gestures. By holding these for several frames, they are not missed, despite the rapid action that is taking place.

Below: An electronic easel enters artwork into a computer for the 'all computer' animation film 'Dilemma'. In the film, computer-generated in-between images create the metamorphosis of a king's head into a knight in armour.

Right: This modern computer animation system—System 4 from the Computer Image Corp, USA—produces animation instantly.

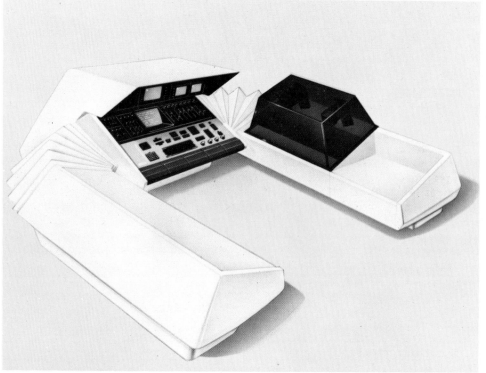

backwards in precisely calculated steps so that the cartoon figure appears to be moving forwards during a sequence.

By allowing a computer to operate the rostrum, a far more precise sequence of movement can be achieved. The result is labour-saving, and gives smoother, more realistic, motion. The device has also opened up the field for complex special effects—which have been used in recent feature films such as 'Star Wars', 'Superman' and their sequels.

Computers are used in animation, however, to do much more complex work—including the actual drawing and animation. But before a computer can work on or store a picture, it has to be translated into computer language. To do this, the image is broken up into a mosaic of tiny squares of tone. Each square or 'pixel' (picture element) is then assigned a number that represents its tone level. Thus the artist's drawing becomes, for the benefit of the computer, a string of numbers.

All the picture information is stored in the computer's memory, which may hold as many as 1.8 million pictures. The advantage of this system is that once a picture is drawn

it need never be drawn again. The computer can be instructed to recall and display any image or sequence of images from its memory, so pictures can be reused in the same form or very easily modified.

One computer system (called PAINT) uses a video camera or laser beam to scan a drawing for toning. It then displays drawings for the artist to work on. The animator communicates with the computer via an electrically sensitive drawing board, which is a flat piece of plastic with an embedded mesh of wires. The artist touches the surface with an electronic pen, causing a mark to appear on a video display at a corresponding point. Using the pen, the animator can 'paint' on the screen or give the computer instructions to move or transform an image. He can also select tones and touch areas to be coloured with the electronic pen. The computer will then accurately add colour to any chosen area of the image.

The system will also colour in-between frames once the animator has coloured certain key frames. Completed frames can then be displayed in sequence for checking before the result is recorded on tape or film.

Perhaps the most exciting capability of the computer is its ability to draw models of characters, vehicles, machines and buildings in three dimensional perspective. It can be instructed to draw models viewed from any angle, close up or far away; it can make them look as if they are being illuminated from any chosen angle, from a light source of any colour or intensity desired. With enough computation, the computer models can even be placed in a computer-generated landscape.

There is little fear of computer stifling creativity: indeed, by liberating artists from the drudge of producing thousand upon thousand of similar drawings, we can expect animation films of the future to be more, not less, original and creative.

AQUALUNG

The aqualung (or SCUBA, short for self contained underwater breathing apparatus) is a system which allows a diver to carry his air supply with him when he dives. He is thus freed of any direct links with the surface and has much greater flexibility of movement than the 'helmet' diver, who needs air and safety lines.

The first really dependable aqualung system was developed by Jacques-Yves Cousteau and Emil Gagnan, but their work was preceded, and undoubtedly helped, by many other attempts. As early as 1879, a British designer named H. A. Fleuss demonstrated a diving set in London in which air was carried in a flexible bag on his back. The diver breathed used air back into the bag, but on the way carbon dioxide was removed from it by caustic potash. Since only a small percentage of the oxygen is used in each intake of air, a relatively small reservoir of air could last a long time. The oxygen actually used was replenished from a small cylinder which he also carried, and this allowed him to stay underwater even longer.

The device was suitable only for shallow depths and even then must have made breathing hard work. But it was good enough to enable Fleuss to clear obstructions from a tunnel under the River Severn which flooded in 1880.

The modern aqualung consists of five basic components: a *demand valve* or regulator, which reduces the high pressure air supply to ambient pressures (the same pressure as the water around the diver); the *cylinders*, which contain compressed air; the *harness* which keeps the apparatus in the correct position relative to the diver's body; the *tubes* for air delivery and exhaust; and the *mouthpiece* through which he sucks the air.

The cylinders which carry the air supply are relatively simple. They are made in various sizes, and the divers like them to be buoyant when empty so that they will float to the surface after use. They are painted grey with black and white quarters at the top, a conventional code to show which gas they contain—in this case air. A typical cylinder might contain about 1.7m³ (60cu ft) of air compressed at a pressure of around 3000psi (200 bar).

The deeper the diver goes, the more air he uses for each intake of breath. Thus the depth of a dive has a bearing on its duration. Cylinders which contain enough air for an hour's diving at 10m (33 feet) will support the diver for only 30 minutes at 300m (100 feet). Other variations in the time scale arise because the diver needs more air the harder he works, or when he is cold.

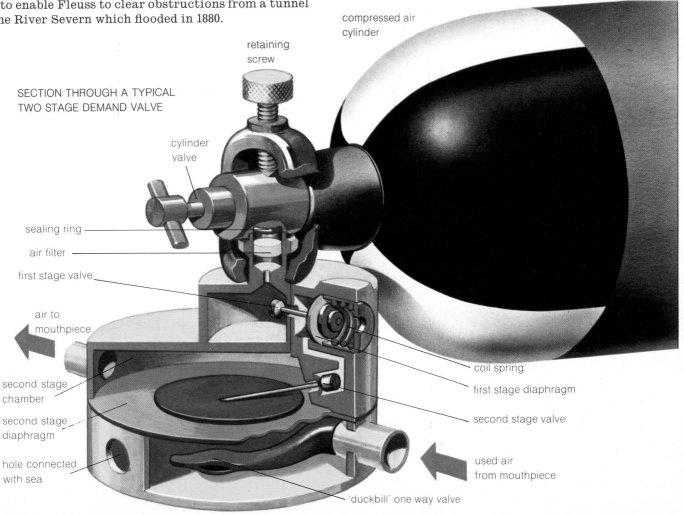

SECTION THROUGH A TYPICAL TWO STAGE DEMAND VALVE

retaining screw

compressed air cylinder

cylinder valve

sealing ring

air filter

first stage valve

air to mouthpiece

second stage chamber

second stage diaphragm

hole connected with sea

coil spring

first stage diaphragm

second stage valve

used air from mouthpiece

'duckbill' one way valve

The harness is usually of nylon or cotton webbing with steel bands. Its task is not only to hold the cylinder in place on the diver's back, but also to make sure that the demand valve is as near the centre of ambient pressure in his lungs as possible.

Demand value

Vital though the other components are, it is the demand valve which is the most complex and ingenious. The human body has evolved on dry land and therefore is designed to withstand the kind of pressures exerted on it by the air (one 'atmosphere' at sea level). Water, though, is much more dense than air, and the diver's body is subjected to much greater pressure in water than on land. The deeper he goes, the stronger these pressures become. Seawater exerts a pressure of two atmospheres (twice the pressure at sea level) at 10m (33 feet) and this increases by one atmosphere for every additional 10m of depth.

The body consists largely of solids and liquids that are virtually incompressible, even under very great pressures. But it also contains cavities which are filled with air—the lungs, sinuses, inner ear and stomach—all of which connect with the human breathing, or respiratory, system. If the air breathed in is not at the same pressure as the water around the body, these cavities will be forced to contract. As air is easily compressed, breathing will become hard work even at relatively shallow depths, and if the diver goes deep enough, the pressure will crush the cavities flat and kill him.

The job of the demand valve is to see that these problems do not occur. The simplest form is a circular box connected on one side to the outlet of the high pressure air cylinder. The other side is open to allow seawater to enter, but the water does not flood the box—it is stopped by a rubber diaphragm inside the open face. The diaphragm is connected within the box to a valve controlling the air supply from the cylinder, and when the diaphragm is pressed inwards that valve is opened.

The external pressure, whether atmospheric or that of the sea, pushes the diaphragm in. This opens the air valve until just enough air has been let in on the cylinder side to balance the pressure. The pressures on both sides of the valve are then equal, so the air the diver breathes must be at ambient pressure.

When the diver breathes this air, he creates a partial vacuum inside the air chamber. The outside pressure opens the valve again, and the process continues.

This single stage design has disadvantages that the high pressures to be controlled by the valve put a heavy strain on it. A modification to overcome this is the two stage type, in which the first stage brings the air down from around 200 bar (3000psi) to about 7 bar (100psi) above ambient pressure by means of a valve acting against pre-set spring pressure. The air can be brought to ambient pressure by a device similar to the single stage type.

In the split-stage or single hose type, the first stage is mounted on the cylinder itself, which is connected by a single small-diameter pressure hose to the face mask, where the second demand valve stage is located.

The aqualung diver's stay under water is limited by the capacity of the air cylinders he carries. Nevertheless, the dive times afforded by the aqualung have made it an important tool in underwater science and technology, and in search and rescue operations, while it has opened the way for diving as a sport throughout the world.

Three stages in the working of the valve. Air at higher pressure than the water is shown in red, at roughly the same pressure in orange and at lower pressure in yellow. The diver opens the second stage valve by inhaling and so lowering the pressure in the second stage chamber. The first stage refills itself automatically. Exhaled air is released nearby in order to equalize its pressure with the incoming air.

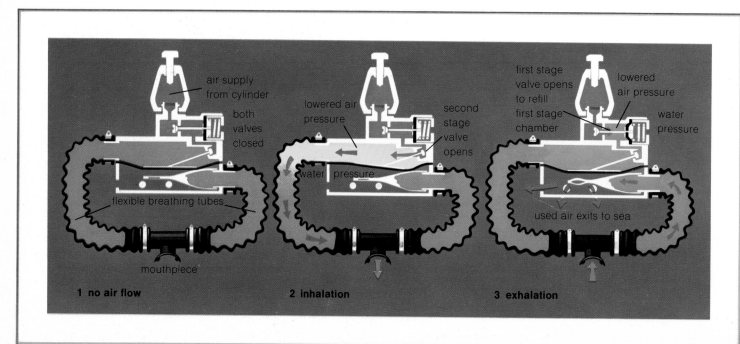

BOWLING ALLEY

The history of bowling can be traced back 7000 years. The grave of an Egyptian child, estimated to have died about 5200 BC, revealed implements similar to those used in the modern game. Modern bowling is played indoors on specially constructed and highly mechanized bowling alleys.

A bowling alley is a long, narrow lane 18.3m (60ft) long and between 104 and 107cm (41 and 42 inches) wide. Ten wooden pins, similar to the skittle used in English ninepins, stand at one end of the lane and the player stands at the other. The player bowls the ball down the lane towards the pins, attempting to knock down as many as possible.

The game is broken down into ten *frames*. A frame consists of one or two attempts by the bowler to knock down as many pins as possible. If all ten pins are knocked down in the *first* attempt or ball of a frame, this is termed a *strike* and no further attempts are made during that frame.

Originally the knocked-down pins were reset by a 'pin boy'; in the 1940s, the first automatic *pinspotter* was developed. It was a cumbersome device, operating on a series of bicycle chains and suction cups. In contrast, the latest automatic pinspotter is a complex combination of electronic and mechanical controls.

Pinspotting is not simply a matter of just replacing knocked-down pins: the pinspotter has to recognize at least four different situations, and carry out different functions for each of them. Computer control is necessary for all this. The four main situations are:

● After the first ball of a frame (if the player has not bowled a strike), the pinspotter has to remove any knocked-down pins (called *dead wood*) from the lane and return the ball to the player.

● After the second ball of a frame, the pinspotter must remove all standing and knocked-down pins and reset, or *spot*, a new set of ten pins for the next bowler, and again return the ball to the player.

● After a strike, the pinspotter must be able to sense whether all the pins have been knocked down on the first ball of a frame, remove them if this is the case and then reset the pins

The various stages in the automatic return of the balls to the bowler by means of the ball lift. This consists of a continuously moving belt and underlane track system.

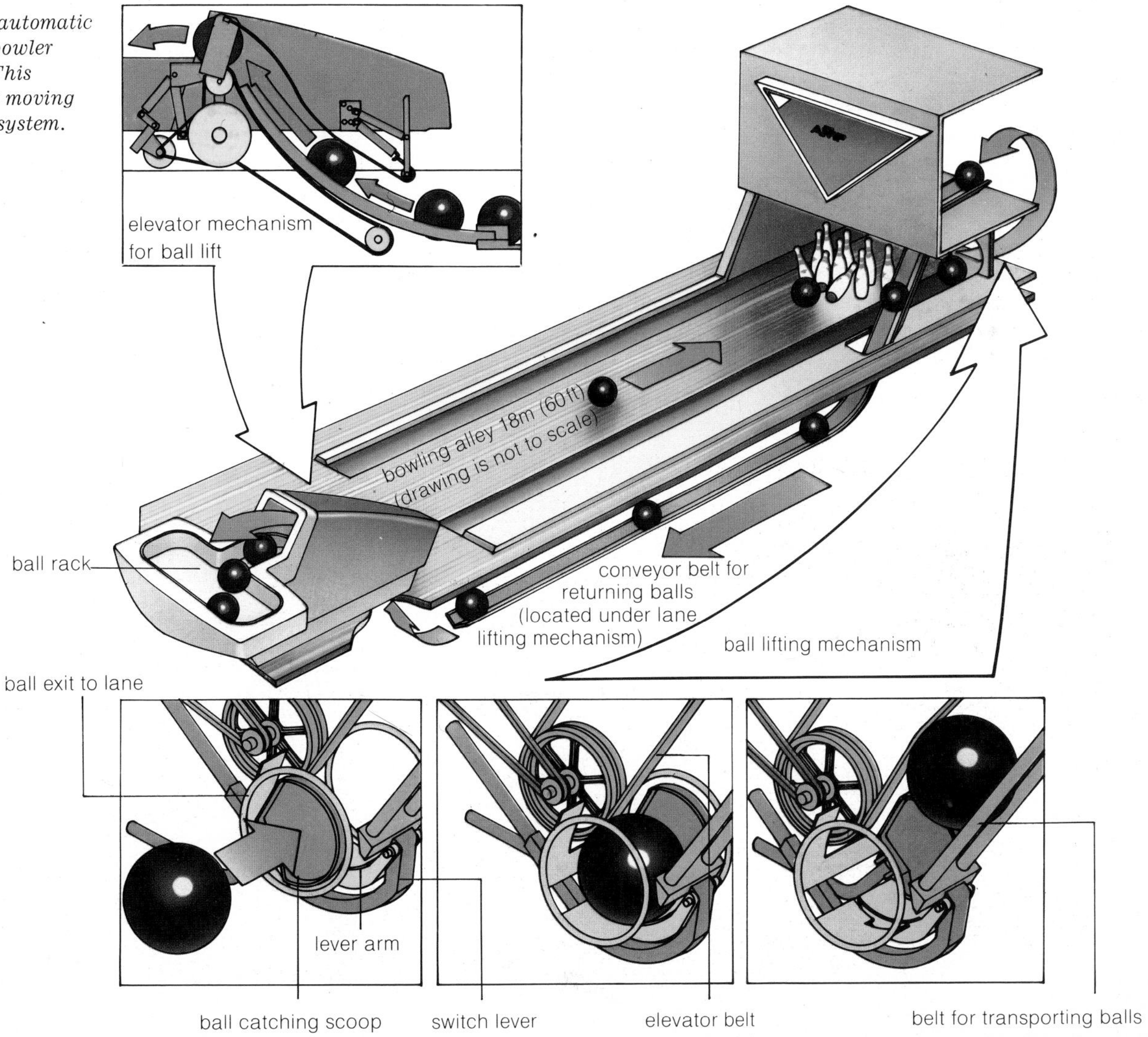

elevator mechanism for ball lift

bowling alley 18m (60ft) (drawing is not to scale)

ball rack

ball exit to lane

conveyor belt for returning balls (located under lane lifting mechanism)

ball lifting mechanism

lever arm

ball catching scoop

switch lever

elevator belt

belt for transporting balls

Right: The automatic pinspotter lowering the pins into their triangular formation ready for the first ball of the frame. The head pin stands at the apex of the triangle, followed by rows of 2, 3 and 4 pins.

Below: View from behind the pinspotter which has just reset the pins. Before the introduction of the automatic machine, a pin boy had the tedious task of either removing fallen pins or resettling them.

for the next frame. (For a strike on the *second* ball of a frame, a pinspotter proceeds as for any second ball—no special actions are necessary.)

● After a *foot foul* (when a player bowls too close to the pins). The pinspotter has to detect a foot foul committed by the bowler as he releases the ball. If this occurs, the bowler loses the benefit of the pins he has knocked down during that delivery.

The pinspotter in action

In all four situations the results are indicated on the display panel (*masking unit*), showing the number of pins standing, the number of pins knocked down, a strike (if delivered), or a foul (if committed). The pinspotter also has to calculate the scores, and display these, and get the bowling ball returned to the player.

In the first situation, when the ball hits the cushion plank of the machine, it activates a microswitch which starts the pinspotter cycle. A bar, the *sweep*, descends across the width of the lane. Its job is to eventually remove the dead wood from the lane but at this stage it guards the pin respotter from any stray balls. After a sufficient time delay to allow for wobbling pins to fall, a gripper device (the *table*) comes down and its 'fingers' close on the neck of any standing bowling pins. The pins are then raised from the lane, allowing the sweep to remove dead wood.

The pins are carried from the lane by an endless belt into the *pin elevator wheel*—a large metal saucer with pockets around the edge to hold the pins. From there the pins are elevated to the *distributor*, which delivers them to the *bin assembly*—ten pockets arranged in a triangle—where they are stored until needed.

After the sweep has removed the knocked down pins on the first ball, the table descends a second time, replacing the pins in the spot from which they were picked up.

By this time the ball has been returned to the bowler by means of the *ball lift* which consists of a continuously moving belt and underlane track system.

The display panel has recorded the results of the first ball and this leads to the next situation, the second delivery of a frame. When this ball is bowled, the same procedure as for the first ball is followed except there is no gripping action from the table and the sweep removes all remaining pins, whether standing or fallen. The table is then lowered and spots a complete new set of ten pins in readiness for the next bowler on a new frame. The score is again indicated on the display panel.

In the third situation, a strike, all ten pins are knocked down with the first ball. The gripping mechanism senses that there are no pins standing and the sweep is brought down to remove all ten fallen pins. The strike is recorded on the display panel with the strike symbol 'X' while the table is spotting a new set of ten pins for the next frame.

In the final situation, a foot foul, the bowler has stepped over the foul line which is embedded in the lane. This action breaks the light beam to a photoelectric cell of the foul detector located on the foul line. This device then passes an electrical signal to the pinspotter which goes into a simulated strike cycle, sweeping away all the pins. The foul is then recorded on the display panel.

HANG GLIDING

Hang gliding, once the sport of a few daredevils prepared to leap from the hills on short and dangerous flights, has progressed to a sophisticated form of sporting aviation. The modern hang glider is an elegant craft capable of flights which cover distances of over 160km (100 miles), make height gains of over 3300m (10,000ft), and last many hours.

Major advances in hang glider technology have produced improvements in performance, controllability and co-ordination. The result is that the efficiency and steerability of today's hang glider compare very favourably with conventional fixed-wing aircraft. The most obvious difference in the means of achieving control is that the hang glider has flexible wings and is steered by the pilot shifting his weight.

To turn his machine to the left, the hang glider pilot shifts his weight to the left, and by shifting to the right, he turns to the right. For pitch control (fore and aft control) the pilot either pulls himself forward, thus lowering the angle of attack to speed the glider up, or pushes himself rearward in order to increase the angle of attack, and so slow the glider down as much as he wishes.

By adopting a prone (near horizontal) flying position, pilots have produced a reduction of pilot drag in the order of 50–60 per cent. (Pilot drag is produced by the rapid passage of the pilot's body through the air.)

Harnesses have also undergone a series of design changes. The latest types are as streamlined as possible, and cocoon the pilot from shoulders to feet. They are also padded, and given protection from the freezing temperatures encountered at the extreme altitudes now commonly achieved.

Most glider designs are derivatives of the early Rogallo Delta shape (named after Dr Rogallo, a NASA scientist), and are tailless. But faster flight and greater efficiency are now being achieved by making the wings longer and more slender —this is called increasing the *aspect ratio*.

However, gains in efficiency are not made without some penalties being paid. A high aspect ratio wing can weigh 25 per cent more than a low aspect ratio wing. It also tends to be much less forgiving of lapses in the pilot's concentration. With a long, high aspect ratio wing, the wing tips are separated by a considerable distance, and each may be subject to a different airflow—the difference may be enough to crash the glider. Lower aspect ratio wings, although not as efficient, have a more acceptable all-round performance, and so are better for the less expert flyers.

Space-age materials

Fifty years ago, hang gliding would hardly have been possible —simply because materials with the necessary combination of strength and lightness had not yet been developed. But in recent years hang glider design has progressed dramatically, thanks to advanced materials originally produced for yacht construction, and aluminium tubing developed for television aerials.

These new materials have stimulated such rapid progress in hang glider design that a new generation of machines has emerged demanding even more specialized materials and components. As a result, the attention of designers is turning to composite materials, developed initially for use in

Above: An assistant trains a new pupil. Tether lines prevent the student from stalling or rolling the machine into the ground. A reasonably strong wind is required to keep the craft airborne, and to stop it drifting down the hill. Only modest skills are needed to fly a hang glider—to progress from complete novice to solo flying takes about six hours. Training schools are fairly plentiful.

Right: Thermal soaring on rising air currents.

space. At present these materials are expensive, and their inclusion would probably put the cost of a hang glider very close to that of a light aircraft.

Powered glider

One of the major limitations to hang gliding is its dependence on suitable weather and another is the need for a suitable hill site. Therefore, it is hardly surprising that enthusiasts have given a great deal of thought to the possibilities of motorizing their craft. A powered hang glider has obvious advantages over a non-powered one, but it also has a number of advantages over light aircraft. First, it needs much less space for storage than a small plane. The hang glider can be folded down in about 15 minutes (by one person) to two small units—the folded wing and its supports forming a neat package 20cm (8in) diameter by 7m (21 feet), plus the engine and propellor assembly (which is about the size of an elongated outboard motor).

Maintenance costs of a motorized hang glider are modest, and fuel consumption is also reasonable—an hour's flight will use only about 4.5l (1 gallon) of two-stroke petrol/oil mixture.

The airspeed of a motorized hang glider is, of course, considerably lower than that of today's light aircraft. This means that a foot-launched motor hang glider does not need a specially prepared airstrip. Low airspeed is also a positive advantage for certain non-sporting applications: searching for missing persons, for instance.

Fitting a power pack to a hang glider does not prevent it from being flown as an unpowered machine—so the best of both worlds can be enjoyed. For example, the glider can fly under power to an altitude sufficient for unpowered thermal-ling flight whose duration will then depend only on the skill of the pilot and the prevailing weather.

But foot-launched powered hang gliders have their disadvantages, such as the problem of the *thrust-drag couple*. This occurs when the centre of thrust of the engine pushing the glider forwards is higher than the centre of drag (which is pulling the glider back). This can, in certain circumstances, produce a turning force tending to pitch the glider over in flight—a highly dangerous instability. These problems and the associated one of the limitations imposed by age or less than peak physical fitness of the pilot were solved by the introduction of a *wheeled undercarriage* such as the Tri-cart. Indeed, this new development has brought powered hang gliding within the reach of an increasing number of people. The use of a Tri-cart, though, does require a small take-off strip about 20m (66ft) long. However, it does not demand much more storage space than a normal powered hang glider.

The introduction of Tri-carts to powered hang gliding has opened up a new range of applications for hang gliders. The cost of flying per hour is a good deal lower than for light aircraft or helicopters and Tri-carts are replacing these machines for aerial photography and military use. The British Ministry of Agriculture is also investigating the use of Tri-carts for spraying, chiefly because of their low cost. There is no doubt that the hang glider/powered Tri-cart combination is one with immense potential. It is easy to see its usefulness to farmers in isolated areas, or widespread bush country, where livestock may move over long distances. Indeed, instead of being just enthusiasts' or specialists' machines, the powered hang glider may one day be a common means of personal transport.

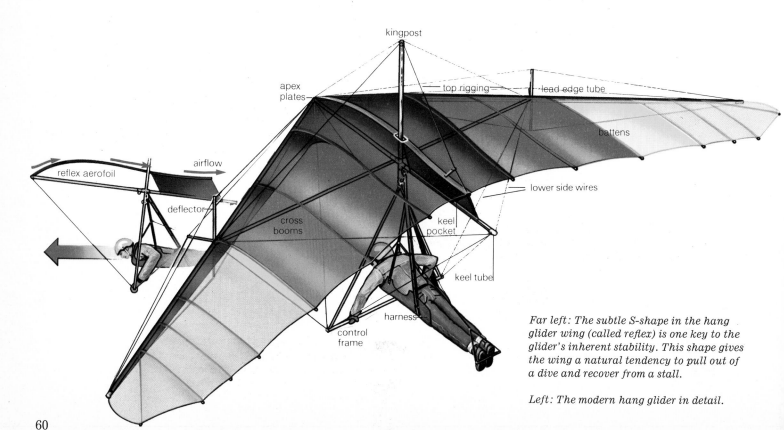

Far left: The subtle S-shape in the hang glider wing (called reflex) is one key to the glider's inherent stability. This shape gives the wing a natural tendency to pull out of a dive and recover from a stall.

Left: The modern hang glider in detail.

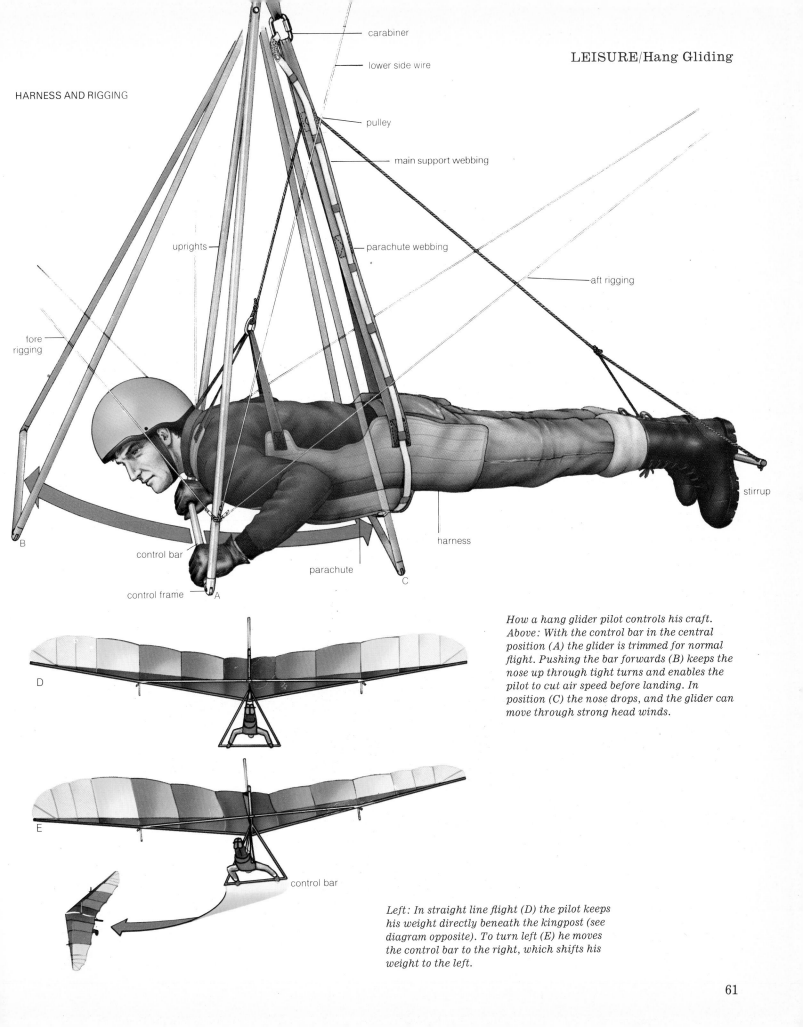

HARNESS AND RIGGING

carabiner

lower side wire

pulley

main support webbing

uprights

parachute webbing

aft rigging

fore rigging

B

control bar

control frame   A

parachute

harness

stirrup

C

D

E

control bar

How a hang glider pilot controls his craft. Above: With the control bar in the central position (A) the glider is trimmed for normal flight. Pushing the bar forwards (B) keeps the nose up through tight turns and enables the pilot to cut air speed before landing. In position (C) the nose drops, and the glider can move through strong head winds.

Left: In straight line flight (D) the pilot keeps his weight directly beneath the kingpost (see diagram opposite). To turn left (E) he moves the control bar to the right, which shifts his weight to the left.

INSTANT PICTURE CAMERAS

Since the first instant camera appeared, about 80 million have been sold around the world, the great majority produced by Polaroid. Colour prints can now be produced instantly, with the image appearing from a blank sheet in seconds—without the need for the darkrooms, negatives, printing papers and sequences of chemical baths that are so much a part of the normal photographic process.

The first instant camera, developed in 1948 by American scientist and chairman of the Polaroid corporation, Edwin Land, was similar in appearance to the folding cameras popular at the time and was a roll-film model. The prints were sepia in colour, rather than the more generally preferred black and white, and the film was insensitive to red light. However, the time taken for the entire developing and printing process was only 60 seconds. For the first time, the photographer could point the camera, adjust the simple exposure controls, take the picture, and then see the results almost immediately.

Land's attempt to duplicate the normal photographic procedure inside the camera itself obviously required a great deal of ingenuity. The film included the 'negative' and the 'printing' paper sandwiched together with all the chemicals required to process them.

Wet processing

When the instant camera's shutter is fired an 'imprinted' negative is formed in the same way as with the conventional film, then the transparent negative and the print paper are drawn, either manually or automatically, between rollers which force them together. As the two materials pass through the rollers, a pod of chemicals attached to the negative is burst and spread evenly between negative and

This Polaroid SX-70 camera folds flat for easy carrying, and contains a host of automatic features. Exposure is automatic, with the whole process governed by solid-state electronics: the exposure is terminated when a photoelectric cell registers that the correct quantity of light has passed through the lens. Focusing is automatic too—see opposite.

THE POLAROID SX-70 LAND CAMERA

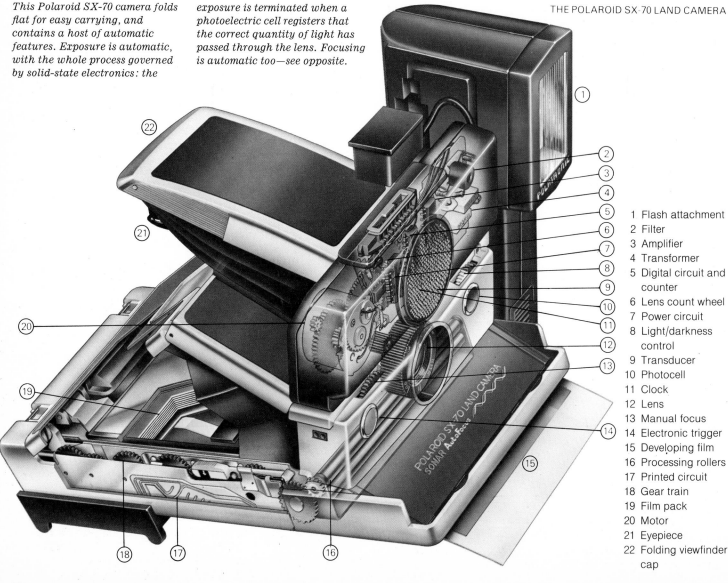

1 Flash attachment
2 Filter
3 Amplifier
4 Transformer
5 Digital circuit and counter
6 Lens count wheel
7 Power circuit
8 Light/darkness control
9 Transducer
10 Photocell
11 Clock
12 Lens
13 Manual focus
14 Electronic trigger
15 Developing film
16 Processing rollers
17 Printed circuit
18 Gear train
19 Film pack
20 Motor
21 Eyepiece
22 Folding viewfinder cap

positive. Initially, this acts as a developer on the negative, but instead of washing away the unexposed parts of the image and leaving them clear, as in a conventional negative, it transfers them as a dark dye across to the print paper which is now held in contact with the film negative.

The result is a positive image on the paper, exactly the same size as the negative. After either ten or 20 seconds, depending on the type of film, the negative is peeled away and discarded, leaving a print on the paper, which develops a little further over the next five minutes, as the paper dries completely.

Despite its effectiveness, instant processing does have a fundamental drawback in relation to conventional photography. Because the negative is contact printed, it has to be quite large for an acceptable size of print to be produced

—in turn, cameras have to be quite large. However, improvements in camera and film technology have contributed to a progressive reduction in size.

Camera body

The solution has been an entirely new camera using perhaps the most advanced of all films: SX-70. The camera is a folding model to achieve the great lens-to-film distance that is necessary for the enlargement of the image, in a much smaller space. An ingenious configuration of mirrors makes the camera, in effect, a single-lens reflex type. The photographer actually sees the image formed by the lens inside the camera before it is recorded onto film. This has many advantages but in the main it makes perfectly accurate composition possible, and allows accurate focusing as well.

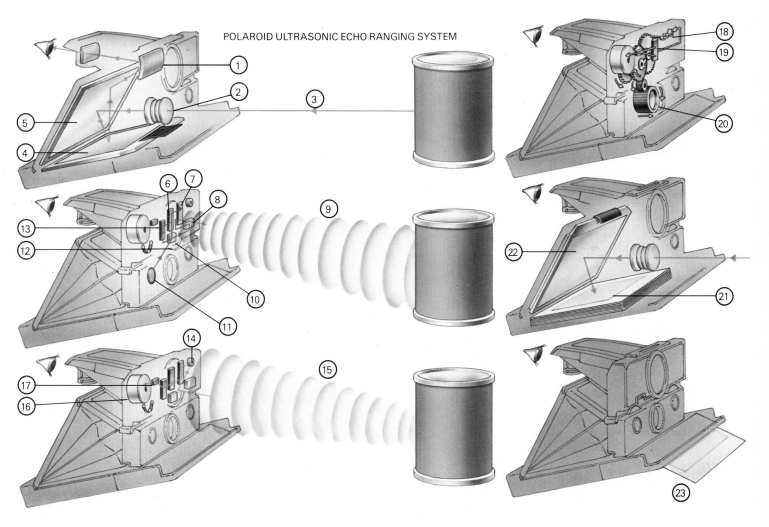

POLAROID ULTRASONIC ECHO RANGING SYSTEM

The automatic focusing system on the Polaroid SX-70 makes use of ultrasonic 'echo-sounding' techniques. The camera shutter button is an electronic trigger, which switches on amplifier and power circuits that send out an ultrasonic beam; this bounces off *the subject and returns to the camera. The time taken for the beam to return, which depends on the distance to the subject, is measured. The lens is then driven by motor to the correct focus position, and the picture is taken.*

1 Aspheric mirror	8 Transformer	16 Motor
2 Four-element lens	9 Ultrasonic beam	17 Motor drive
3 Light	10 Clock	18 Solenoid
4 Fresnel mirror	11 Electronic trigger	19 Counter
5 Mirror	12 Transducer	20 Lens
6 Digital circuit and counter	13 Power circuit	21 Exposed negative
7 Amplifier	14 Filter	22 Trapezoidal mirror
	15 Reflected beam	23 Photograph

With a dry process instant picture film, you can watch the picture develop before your eyes.

Dry process films

Perhaps the most remarkable feature of the SX-70 system is the colour film itself. Using metallized dyes, the process is a dry one, because the developing and printing chemicals are sealed within the film covering. There is no messy negative to peel away and discard; and the picture develops as you watch. Kodak have also introduced an instant camera system, with a colour film similar to SX-70 (Kodak Instant Print).

A look at the working of the SX-70 colour film shows how advanced chemistry is combined with high-technology camera mechanics. The film is supplied in packs of ten which simply drop into the back of the camera. When the shutter is fired, exposing the film, the print is automatically wound out of the camera. It consists of a rectangle of white plastic on which the colour image gradually begins to appear, taking up to four minutes for complete development. Any normal film would be immediately ruined if allowed into daylight in this way before the processing procedures are complete.

To achieve this the film consists of a complex sandwich of a number of physical layers, most of which are coated on the two plastic sheets of which the film is made up. At the bottom of the sandwich is a dark plastic base on which is coated the many layers which will go to make up the negative and supply the dyes for the print. At the top of the sandwich is a clear, transparent plastic layer which forms the protective surface of the print. On its underside is coated the image receiving layer.

Between the multi-coated base and the clear surface layer there is a weak bond. When the film is exposed and is forced out of the camera, it passes through two rollers. As with the earlier instant film, a pod of chemicals in the film is burst by the rollers and chemicals are forced evenly between the layers. In this case, it is a thick, milky substance which, as it is spread through the interior of the film sandwich, prevents light from reaching the light sensitive layers lower in the sandwich. At the same time, it acts as a developer and sets in motion the processing of the film.

Apart from space layers, the negative portion consists of six active layers. Three of these are coated with light-sensitive silver halide and three contain red, blue or yellow dyes. The layers are so placed that they act as filters for the layers below so that, as the light passes through them, colours are progressively filtered out and each layer responds to one particular colour. During exposure, three negative 'imprints' are thus formed, one each for red, blue and green light. After the developer is added, dyes from the other three coated layers are released and move freely through the developer mixture. When they meet an exposed part of the negative of the appropriate colour, however, they are immediately trapped and neutralized. Those that are not seized in this way, still mixed with the white developer, are trapped on the image forming layer beneath the surface of the print.

Two things happen next. First, the coloured pigments mixed with the white pigment begin to form a positive image and, second, the opaque colouring elements in the reagent mixture are gradually cleared by the materials released in the development process. After about a minute, the opaque white 'barrier' becomes clear in accordance with the level of chemical activity beneath it, revealing the picture which rapidly becomes fast on the image receiving layer. Another, slower working layer now fixes the activity and the image becomes permanent.

Instant photography users

While the greatest market for instant photography has proved to be in the 'snap-shot' market, the development of the technology has been extended into diverse fields. There is a surprisingly high use of instant equipment by professional photographers, in both the creative and the industrial fields. Special Polaroid film backs can be fitted to a number of professional cameras and these are used extensively for planning the composition and complex lighting of a shot. The photographer simply fits the Polaroid back to his or her camera and checks the results of his planning on instant film before switching to conventional film when satisfied. There are also industrial photographic uses, with professional instant cameras available which can produce high quality records of engineering works in progress, for example. But the question is—how will instant photography stand up to the video revolution?

MODEL RAILWAYS

Playing with trains is something that few children (and even fewer adults) can resist. Yet, despite the connotations of 'toys', model trains are something to be taken very seriously —even the train sets produced primarily for children are highly sophisticated miniature examples of the real thing.

The most fascinating thing about a model train layout to most people is watching the trains moving—stopping and starting in stations and at signals, shunting in sidings, and carrying out all the complex manoeuvres that can take place on a full-sized train system. Control for model train movements has developed rapidly over the last few years.

Speed control

The basic control of small scale electrically powered railway locomotive models is made a straightforward matter by the use of permanent magnet motors which reverse their direction of rotation when the DC supply is reversed in polarity— so the train can easily go forwards and backwards. Speed control is arranged by adjusting the voltage, usually from around 4V to the full output of the unit which, although rated at 12V, is frequently nearer 16V.

For many years two methods of speed control were favoured—the *variable resistance* and the *variable transformer*. The former is more flexible in use but does not give particularly good speed control with the motors usually used. The variable transformer is a true voltage control device but, since it has to be incorporated in the power unit, it is practicable only when a manufacturer can produce the specialized transformers required at a reasonable price and is outside the scope of the amateur.

Transistorized controllers

The development of power transistors ushered in a new type of control. The initial transistorized controls used a simple *emitter follower* amplifier which allowed true voltage control. Furthermore, the addition of delay circuits provided steady acceleration to a pre-set speed and the possibility of simulat-

ing the effect of a brake. Later, the more sophisticated *feedback controller* was developed. This senses the voltage generated by the motor in the idling mode between pulses and adjusts the output of the controller to maintain a constant speed. Its main advantage was that it was much simpler to operate.

The electricity supply is usually fed to the locomotives through the rails on which they stand so it follows that, unless something is done, all the locomotives on a particular track will move at once. In practice, there are several ways around this problem. The simplest is to provide isolation of portions of the layout so that locomotives standing on these sections cannot move. In many cases the turnouts (points) are arranged to isolate sidings automatically and the main track is divided into different sections by insulating gaps and fed through on-off switches. This elementary type of control is remarkably effective since one operator can, in practice, drive only one train at a time anyway.

For large circuits, a more sophisticated version of this control, called *block sectionalizing*, can be used. Here, the various sections are controlled by relays. As the locomotive passes a fixed point, a relay isolates the section immediately behind the train: on passing a second trip at least one train's length ahead, the first relay is restored to normal and that section re-energized, while the section now immediately behind the train is isolated in turn. With three relays, two trains can move alternately around a circuit. Four blocks will ensure that both trains can move at once but five or six are preferable for continuous operation.

Command control

However, the trains are not under independent control. For this, one needs the latest development, *command control*. Here each locomotive is fitted with a module which operates

Below left: Model of Dutch Loco 1306 shown on ready-ballasted track with model catenary wire and wire mast system; the model can run on overhead or on track powered systems.

Below: Model of the County of Bedford (an old British steam train). It can be operated by a command control system. Other special features include a realistic smoking funnel.

only when the coded signal initiated by the controller is received. Although command control is by no means a new idea (such a system was commercially developed in the 1960s) the large size of the locomotive modules and, in many cases, the small number of control channels available made the scheme unattractive. However, in recent years, the problem of miniaturizing the locomotive module has been solved while at the same time the number of channels has been increased to 16. One system utilizes frequency control. Another employs a digital command signal, and in addition to the 16 locomotive channels, can provide from the same control unit 99 control circuits for points and signals, all operated by linking them to the track via a special module.

The disadvantage of command control is that trains have to travel in the same direction at much the same speed. However, the advantage is that no sectionalizing is required and all locomotives remain stationary until called upon to move. It is at its best on compact complex layouts, where the movements seen around a busy station in real life can be simulated.

As with full sized railways, model railways are ideally suited for computer control, (and have provided the basis for at least one university student's thesis). However, the ultimate enjoyment of building a model railway lies in having complete control over its operations—and a computer could easily remove such fun.

TRACK LAYOUT AND CONTROL UNIT

Left: Zero 1 Micromimic Display System; the track layout is reproduced on the display panel, which also indicates the number and setting of the points. A later development might be the use of a light pen to alter settings. A British Rail signal box (above right) has a similar display—but bigger!

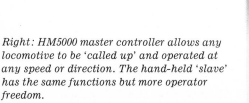

Right: HM5000 master controller allows any locomotive to be 'called up' and operated at any speed or direction. The hand-held 'slave' has the same functions but more operator freedom.

ROLLER COASTERS

The sophisticated designs of modern roller coasters offer visitors to amusement parks throughout the world the nearest thing to fear of imminent death in total safety. The all-steel structures carry the passengers in Teflon-wheeled steel and fibreglass cars around tortuous loops and abrupt bends at high speed.

Old-style roller coasters were complex structures of wood and steel scaffolding. Most have now been pulled down, because of the difficulty of maintaining the rapidly deteriorating timbers and rusting steel. Modern coasters are of all-steel construction, and often virtually self supporting and of modular design, so that building time on site is minimal and, just as important, the ride can be quickly dismantled when the public no longer finds it thrilling enough.

Probably the most advanced of the new-style coasters can be found in the World of Fun Amusement Park in the cattle centre of Kansas City, Missouri, USA. Claimed to be the most advanced ride ever built, this roller coaster, called the *Orient Express*, incorporates two complete loops, followed by the novelty of a turn about the axis of 360 degrees as the screaming passengers hurtle earthwards.

The train, consisting of six deluxe steel-trimmed cars, starts at ground level and begins its motor-driven climb at a modest speed up a 25 degrees slope to a height of 78.5m (260ft). At this point the train is carried around a 32 degrees bend and, with disturbing suddenness, it is swept downwards at an enormous velocity, reaching a speed of 96km per hour (60mph) in just over two seconds. At the end of the steep drop the cars reach a basin-shaped curve, at the bottom of which the passengers are subjected to a downward pull more than three-and-a-half times that of gravity. Throughout the ride, such 'G' forces affect the passengers' features like a body massage—the ride is designed to give the feel of maximum bodily dislocation in total safety. Mouths opened to scream

The complex framework of tubular steel, which provides a safe support for the steel rails, is shaped into elegant designs.

are quickly forced shut again as the train moves up and out of the curve and is thrown into the first loop. Passengers travel momentarily upside down as they move at extreme speeds (for the size of the train) round a loop with a diameter of 14.3m (48ft). The second loop follows the first with alarming rapidity, and in both cases the inverted heads of the shrieking passengers are travelling at a height of almost 24m (80ft) from the ground.

The main body of the *Orient Express* is made of several lengths of tubular steel piping with walls 15mm (⅝in) thick and a diameter of 300mm (1ft). Welded to this main member, which coils around the whole circuit, are boxed steel brackets to which parallel lengths of narrower gauge tubular steel are fixed as rails. The cars are made out of a very durable moulded fibreglass, made to appear more glamorous by steel surrounds. There are two upper wheels and one lower runner on each corner of the car. They are fitted tightly on either side of the steel rails and coated with Teflon to reduce friction.

The Revolution

The Blackpool (England) coaster, *The Revolution*, begins with a crank lift ascent to a uniquely constructed station and launching platform made of a steel lattice framework which houses the propulsion unit at one end. Here, passengers enter the four-car train and fit adjustable double-locking padded safety bars over their shoulders. These bars mainly serve the function of psychological reassurance, as the negative gravity force of the ride is more than enough to keep the passengers in their seats. The chance of a breakdown, which may temporarily suspend the passengers upside down, is rendered minimal by a computer terminal control unit (the Solid State Westinghouse numerological control panel) which monitors almost 18.5km (11½ miles) of wiring.

To set this ride in motion, a 200-volt DC motor releases an air-compressed launching hammer to thrust 125 horsepower into the back of the train. The impact throws the train forwards along the short horizontal section. It then travels down a slope, reaching a speed of 43km per hour (27mph) at the bottom before being hurtled into the first loop. The loop itself is only 15m (50ft) long, so its diameter is significantly smaller than its Kansas City counterpart, making the changes in gravitational and negative gravitational pull even more dramatic than on the *Orient Express*.

The wheels of each car grip the 125mm (49in) diameter tubular steel rails by means of a locking bar. The system of 1.2m (4ft) wide rails, supported by a main 300mm (1ft) diameter pipe, is basically the same as in the *Orient Express*. Supporting the belly of the loop at either end are shorter sections of thick boxed steel resting on small concrete plinths. This ensures that although the ride is firmly and safely anchored, dismantling it would not prove a great problem. Moreover, the design of the all-steel system can allow for alternative balancing leg supports which make the whole structure easier still to move.

The Looping Racer

Another ride firm, Schwarzkopf, has developed the *Looping Racer*. This is launched by a catapult start above ground level at the beginning of the track in a conventional roofed station. The beauty of the design reaches its peak in the

Right: A modern roller car is made of tough glass fibre reinforced plastics, enhanced by stainless steel trim and fittings. Two pairs of wheels on each side of the car sit on top of the tubular track rails, and two small pairs grip the rails from beneath to reduce vibration and ensure a smooth, safe ride. Wheels are Teflon-coated to reduce friction to the absolute minimum. The car will not start until safety bars have been locked across the passengers' legs. Far right: A masterpiece of elegant design—The Shuttle Loop.

adaptability of the air-compression starting mechanism. The catapult's power can be geared to a number of different site layouts. The momentum of the cars resulting from the starting push is a vitally important design factor in all coasters: the lower the starting mechanism is sited, the less momentum is produced and the greater the power required to thrust the train forward. Schwarzkopf's starting catapult is powerful enough to fling the train out of the station and round the first loop in one go, thus giving sufficient momentum to the train to complete the course, which involves a full circle back to the starting point. The *Looping Racer* reaches a speed of 80km per hour (50mph) within a few seconds of the start, and negotiates an astonishing incline of 60 degrees to climb into the 20m (65ft) diameter loop.

Speed, dramatic effect, and versatility are the hallmarks of the new steel roller coaster—but our love of nostalgia has already persuaded one firm to start building modern versions of the old, creaking, wooden coasters.

STEREO

The world's greatest classical composers were fated to hear their symphonies only rarely, at concerts. To turn on music at will, day or night, reproduced to a standard almost indistinguishable from a live performance, is the gift of hi-fi technology. It has revolutionized popular and classical music making and led to an explosion in the number of people who can now listen to music every day of their lives. New developments, including indestructible discs played not by stylus but by laser beam, are in active progress. The story of hi-fi is one of continuing improvement to a few basic techniques.

A hi-fi system's task is to convert the coded input it receives (from disc or tape or via a radio tuner) into audible vibrations in the air—sound waves.

What is hi-fi?

Hi-fi was a term coined to describe the best available record players, designed to reproduce music with a high degree of fidelity to the original sound at a performance. Today's best hi-fi systems are often better than the sources (disc, tape or radio broadcasts) that they use—so the term has changed its meaning slightly.

In addition, to speak about high-fidelity to an original performance means little when that original may never have existed. In the case of electronic music or music recorded on a multi-track tape recorder, each instrument is recorded separately and the complete work does not exist as a whole until all its elements are assembled in a mixing studio. Indeed, most of today's pop music is recorded this way.

Music can be encoded and preserved in a variety of ways; one of the most popular storage mediums is disc. On disc, music is encoded in a vinyl surface as a wavy groove whose shape corresponds to the waves the microphone receives as sound waves from the original source. The sound of many instruments playing together produces a complex and tortuous groove path that can contain information about the acoustic space in which the recording was made, as well as its tones and harmonies.

Stereo discs

With the right equipment the spatial aspects of the music can be reproduced at home. The simplest way of doing this is stereo, where two speakers are fed with separate signals amplified separately, in order to mesh acoustically in the listening room and create a sound 'image'.

How the human ear perceives stereo is a complicated subject, but what is important is that the input to the system must consist of two separate signals recorded simultaneously by different sets of microphones.

On discs these two signals are encoded in one groove. The movement of one wall of the groove corresponds to one signal and the movement of the other wall corresponds to the second signal.

Before these groove wall patterns can be transformed into sound they must first be converted into corresponding electrical signals, as varying patterns of current; then amplified to a strength at which they can drive a pair of loudspeakers. The speakers convert the electrical signals back into high energy sound waves.

At each stage of conversion and amplification, distortions can be introduced into the signal by the mechanical and electronic process involved. Hi-fi systems have elaborate circuits to correct or prevent these unwanted additions to the signal. It is how these circuits are designed and used that distinguishes a good sound system from a poor one.

The greatest distortions are introduced where one form of signal is converted to another, whether from a mechanical

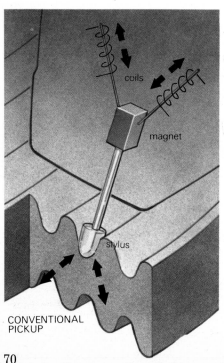

CONVENTIONAL PICKUP

Left: Sound waves are stored on a record in the form of a wavy groove. As this passes underneath the stylus of a pick-up cartridge, it causes the stylus to vibrate. In turn, this causes magnets in the cartridge to vibrate, and this induces a varying electrical current in tiny coils of wire. This electrical representation of the sound wave is passed to an amplifier, and then to the loudspeaker.

Below: A typical magnetic cartridge, showing the stylus, coil and magnet arrangement.

remote control

form to electrical (as in the disc pick-up) or vice versa (as in the loudspeaker). As a result, correct design of the pick-up and loudspeaker is crucial to correct translation of the original signal on the disc.

Pick up cartridge

Most modern pick-ups consist of a tiny magnet linked to an *armature*, at the end of which is a diamond stylus. The stylus traces out the path of the record groove, and so transfers the groove vibrations to the magnet. This magnet is surrounded by very fine coils of electrical wire. As it vibrates and moves in and out of the coil, the magnetic field cutting across the coils changes and causes a current to be introduced in the wires. This is fed to the input of the amplifier and then the loudspeakers. Many subtle distortions can be introduced if the stylus does not exactly follow the shape of the groove as the record revolves. Altering the profile of the stylus can help; the most usual profile found in hi-fi systems is the elliptical.

In order to get two signals out of the one groove, the magnet is linked to two coils, arranged so that movements of one wall of the groove produce an electrical output from just one coil, and movements of the other groove produce an output from just the other coil.

Although good design of magnets and coils minimizes in-trusion of one signal into the other coil, there is always some leakage and this can partly destroy the stereo effect—the 'spaciousness' of reproduction.

Magnetic tape

Storing music as mechanical waves cut into records is not the only possibility. Music can also be encoded onto a magnetic tape as stripes of varying magnetization in an iron oxide film which is coated on an acetate base strip. By running the tape past a coil of wire in a *playback head*, the varying magnetic levels on the tape will induce a correspondingly varying current through the coil of wire.

Large open reel tape recorders, popular before the advent of the music cassette, used wide tape and recorded up to four tracks on the tape. Two tracks in one direction provided the two signals necessary for stereo, two in the other direction enabled the playing time of the tape to be extended. Most open reel tape decks offered a choice of three running speeds, permitting economical recording at low speeds, or better reproduction at higher speeds.

The advent of cassettes, originally intended as quick-load, convenient capsules for dictating machines, effectively ousted the open reel recorder, except in the field of live music recording. The width of a cassette tape, at just one-eighth of an inch, is half that of a standard open reel deck. Crammed

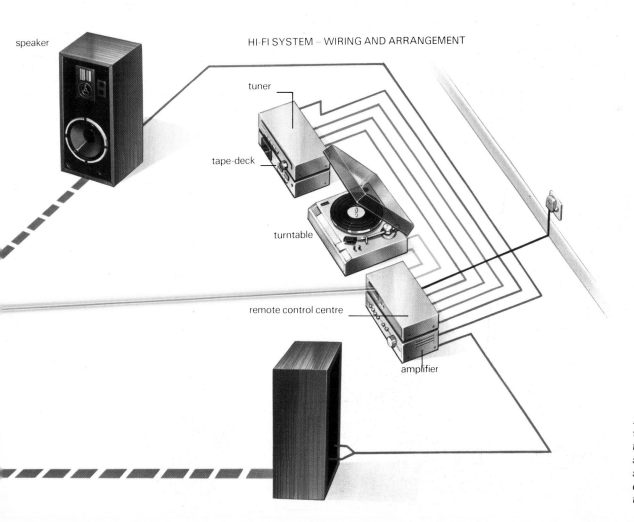

speaker

HI-FI SYSTEM – WIRING AND ARRANGEMENT

tuner

tape-deck

turntable

remote control centre

amplifier

Left: A hi-fi system wired for use. With the most sophisticated systems even the shape of the room has a significant effect on the sound created.

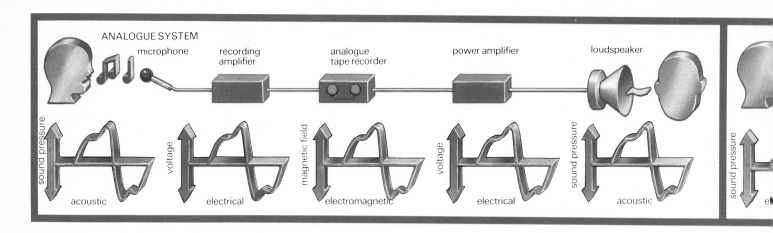

ANALOGUE SYSTEM

microphone — recording amplifier — analogue tape recorder — power amplifier — loudspeaker

sound pressure | voltage | magnetic field | voltage | sound pressure

acoustic — electrical — electromagnetic — electrical — acoustic

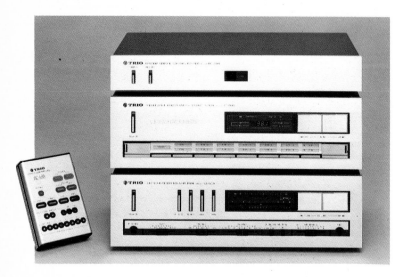

Above: A set of modern hi-fi components, designed to complement each other visually. This system has full remote control.

Right: This speaker has three separate drive units, or cones. The top one handles the higher-most frequencies; the middle one slightly lower frequencies, and the large bottom one the lower-middle and bass frequencies.

on this are four narrow tracks—which presents a challenge to efficient signal transcription.

The narrower a recording track, the bigger, relatively, is the size of the iron oxide particles that carry the magnetization. As a result, there is less available signal relative to the inherent noise on the tape (which comes from the random magnetizations in the oxide particles). In brief, as the track width narrows so the noise on the tape (which is apparent as a hissy background during replay) increases.

Furthermore, cassettes run at a slow speed, 4.76cm/sec (1⅞in/sec), which means that the high frequency response of the tape is limited. Very high frequency signals stripe the tape at intervals which are nearly as small as the size of the oxide particles so proper magnetization cannot take place. It is rare for a cassette deck to have such a poor high-frequency response that the actual high notes of instruments become lost—but all decks lose some of the *harmonics* of a note (to a greater or lesser extent), and the loss of these causes a note to sound dull.

Noise reduction systems

Several systems have been developed to reduce background noise: the most popular is the Dolby system. In essence, a Dolby circuit artificially boosts the high-frequencies in the signal as the music is recorded; then diminishes them on replay so they sound at their proper level. However, diminishing the high frequencies also reduces the level of hiss (which is most apparent at high frequencies). So the music sounds right, but the noise is reduced. The Dolby circuit has other refinements to ensure that the tape hiss is eliminated without affecting the music signals. Other noise reduction circuits operate in a similar fashion.

In addition, to aid reduction of background noise, some manufacturers developed tapes with special oxide layers that become more strongly magnetized than regular tape. Others developed tapes employing oxides of chromium, some of them mixed with iron oxides. The latest development is tape with a pure metal, rather than a metal oxide coating.

Tape technology is still in a state of development. But new tapes and advances made in noise reduction circuits have enabled even middle quality cassette decks to equal most of the earlier open reel decks.

With the growth of the VHF (very high frequency) radio network the demand for high quality radio tuners has in-

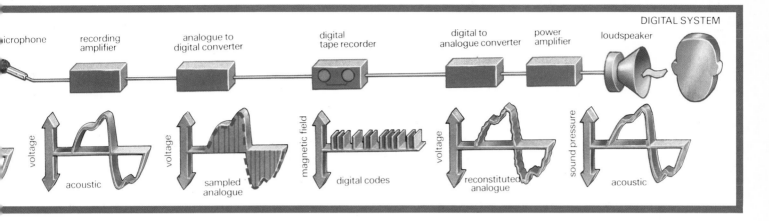

DIGITAL SYSTEM

microphone · recording amplifier · analogue to digital converter · digital tape recorder · digital to analogue converter · power amplifier · loudspeaker

voltage · acoustic · voltage · sampled analogue · magnetic field · digital codes · voltage · reconstituted analogue · sound pressure · acoustic

A comparison of analogue and digital recording techniques. In the analogue system, the sound produced by the singer is represented as a wave form all the way through the recording and reproducing chain—for example, as a continuously varying voltage, or level of magnetisation. In the digital system, however, the sound is treated as a series of numbers. Analogue amplifiers and speakers are still used, so converters from one system to the other are needed.

creased greatly. The tuner is thus a key part of the hi-fi system.

Radio tuner

In a radio broadcast, an electromagnetic radio wave is transmitted at the chosen frequency. The music signal to be broadcast is encoded on this *carrier wave* by changing or *modulating* the carrier in some regular way corresponding to the changes in the music signal waves. There are many ways of modulating the carrier—that chosen for domestic broadcasting on the VHF waveband is called *frequency modulation* (FM). FM's main advantage over other methods is that it is less susceptible to hiss, fading and interference. And the use of the VHF band allows a good high frequency response. With a complex system of further coding it is possible to broadcast in stereo—so all requirements of a good hi-fi system are met.

Amplifier

At the heart of any system is the amplifier. Its function is to detect the appropriate input, whether disc, tape or radio, and amplify the signal received until it is strong enough to drive a pair of loudspeakers. A pre-amplification stage incorporates all the control functions—among them, treble and bass, loudness controls, volume and balance.

Again, the quality of modern amplifiers depends more on refinements to the amplifying process than any other factor. The way in which the amplifier's circuits treat the incoming signals can introduce unwanted or spurious signals which result in a distortion of the sound, but modern amplifiers are totally free from audible distortion.

Loudspeaker

The loudspeaker, like the pick-up cartridge, converts the signal from one form (electrical) to another (mechanical movements of the speaker cone). The simplest loudspeakers consist of a coil of wire between the poles of a magnet. The coil is attached to a cone of paper or plastic which is anchored at the edges of the speaker frame. As the signal current passes through the coil it interacts with the magnetic field between the poles of the magnet and moves the coil in sympathy with the flow of current in the wire. Thus the paper cone vibrates and sound waves are produced.

The situation is complicated by the inability of one size of speaker cone to handle all frequencies of sound equally well. In order to reproduce all the frequencies in the music spectrum, loudspeakers use up to three or four separate drive units, one to handle the bass notes, another the mid-range frequencies (such as voices and most instruments) and one or two units to take the very high frequencies. Inside the speaker is a crossover circuit which divides the incoming signal into the frequency bands most suited to each drive unit.

Like the pick-up cartridge, a loudspeaker can introduce a great deal of distortion—for example through resonances in the material of the speaker cones, which artificially 'colour' the sound output. These distortions are kept to a minimum by making the cone as stiff as possible (by plastic coatings or reinforcing laminates) and by attention to the way in which the edges of the vibrating cone are anchored. In addition, speaker cabinet design is important to the final result. Most loudspeakers feature a solid cabinet, heavily padded on the inside to absorb unwanted sound and prevent the cabinet itself from resonating.

Digital hi-fi

The future of hi-fi (and indeed its present) is being revolutionized by micro-chips and the digital technology associated with them. In the digital age, the encoding processes (to store music on tape or disc, or transmit it through the air) do not form electrical, mechanical or magnetic waves that try to 'look' like the acoustic wave they represent, as in the *analogue* systems described. Digital encoding treats the acoustic wave as a series of numbers—which can be manipulated, stored and transmitted by computer-type methods. Digital processing of a music signal means that many of the distortions introduced by the components described above can be eliminated. Perfect hi-fi may yet be attainable.

VIDEO SYSTEMS

Just as with television, the ability to replay TV pictures in the home is nothing new—Baird sold a video player in the 1930s, along with his primitive television system. But it is only recently that video recording and replay systems have been widely available for the domestic market.

There are two main types of video equipment available. Paralleling their audio counterparts, there are tape-based systems, where you can, in your own home, both record and play back television programs and your own electronically produced home 'movies'. And there are disc-based systems, which by and large are for replay of pre-recorded material only.

Video tape recorder

The main problem with recording video (moving pictures) as opposed to audio (sound) is that much higher frequencies are involved. Audio signals do not extend above about 20kHz (kilohertz) but video signals can go as high as 5000kHz. With a tape-based system, the only way to record such high frequencies is to have the magnetic tape passing the recording head at very fast speeds. Early tape systems did just this, and tape speeds could be as high as 200 inches per second (modern audio cassettes operate at $1\frac{7}{8}$ inches per second). This put a great strain on the mechanical components of the system,

and used up vast amounts of tape, so a way had to be found of slowing the tape speed down while still allowing the speed at which it passed the recording head to remain high. The system developed was known as *helical scanning*.

In helical scanning, the recording head is arranged to lay down recorded tracks across the *width* of the tape, rather than along its length. Then, for every short distance that the tape moves, the total length of recorded track is much longer —thus the tape can move slowly, but the head-to-tape speed is still high. This is relatively simple—the trick comes in arranging the recording head to do all of this, and the answer is to make the recording head in the form of a drum, with the recording gap (the split in the head, which actually causes the magnetic pattern to be laid on the tape) spiral around the outside of it. By spinning the drum round at high speed, as the tape moves, the recording pattern will be in the required form of parallel bars across the width of the tape. Even with this technique, domestic systems find it necessary to reduce the frequency range slightly in order to produce a recorder at a reasonable price—this reduces picture quality a little, but on most domestic programs, it is rarely noticeable.

There is one potential problem—but thanks to the way that a television picture is constructed, it simply does not occur. The potential fault is momentary breaks in the picture, as the recording head reaches the top of one short bit of track, and has to leap down to the bottom of the tape to start the next bit. But because a TV picture is built up frame by frame (see page 11) it is possible to coincide this leap with

A colour video camera adds a new dimension to videorecording. With the film processing stage eliminated, spontaneously taken film can be played back instantly. A reduction in picture quality and the need for powerful lighting are drawbacks, however.

The inside view on Sony's state-of-the-art Betamax C-7 videorecorder. This machine can record up to four programs over a two-week period. On playback, the picture can be slowed down, run frame by frame, frozen or speeded up. An infra-red remote control keyboard is also provided.

the break between frames—and so there is no discernable gap.

A more likely problem is control of the spinning head on replay, so that it synchronizes with the tracks laid down during recording—if the replay head movements are even fractionally out of time, then it will not follow the recorded tracks, and the picture will break up. So all video recorders have a control track running the length of the tape carrying synchronizing pulses.

All current domestic video recorders work in much the same way, but the mechanical details vary between brands so much so that there are half a dozen or more (counting obsolete models) different standards of video recorder all of which are incompatible—that is, tape cassettes designed for use on one system can not be used on any other system.

Video discs

Video disc systems are lagging behind tape ones, but some are now available in America, and should become available in the UK during 1982.

Conventional audio techniques cannot be used for video discs—as with tape, the frequencies involved are just too high. Nor is it really feasible simply to 'stretch' conventional techniques, in the same way that video tape does—a new approach is needed.

Certainly the most technically sophisticated approach is the Philips Laservision. The video signal is coded onto a flat disc in the form of microscopic 'pits', each one less than half a millionth of a metre wide, the length and spacing of the pits carrying the video and audio information. The pits form a spiral track, in roughly the same way as the groove of a conventional audio disc does. But there are many differences in the way that this disc is 'played'. For a start, there is no stylus—instead, the information contained in the pits is read off the disc by a laser. This is used as a light source aimed at the pits as they track round underneath it. The pits, which are metallized, reflect the light in a varying way, and this varying reflection is picked up by photodiodes which produce an electrical signal corresponding to the variations, and hence corresponding to the information coded in the pits.

Accuracy problems

Probably the major problem with this system, as with any video disc system, is the incredible degree of accuracy that is necessary if the laser is to read the information correctly. Because there is no groove in the disc, another method for getting the laser beam to follow the spiral track of the pits has to be found. The laser is driven by a small motor across the disc surface, in the same way as with some 'radial tracking' record-playing arms, but this is not enough to ensure that the laser will be precisely positioned over the pits. So two auxiliary laser beams are used, positioned to scan opposite edges of a pit, plus a part of the adjoining flat surface of the disc. If the main beam is tracking correctly, the signal these side beams sends back is identical, but if the main beam is off-centre, then the signal will be unequal: an error-compensating device checks for any inequality, and corrects positioning constantly.

It is also important to ensure that the main beam focuses exactly on the surface of the pit. Inevitable warps in the disc's surface mean that the focusing position is constantly changing—again, an ingenious system of lenses and photodiodes detects any change in the focus distance, and a moving-coil system alters the position of the laser's focusing lens to provide the correct focusing point. Finally, crystal control ensures that the speed of the rotating disc is always just right.

There are three other strange seeming differences between the Laservision and a conventional audio record player. The laser stylus and all its associated control equipment is mounted underneath the disc, so you have to place the disc in 'upside down'; and the stylus tracks the disc from the inside to the outside, rather than the other way round. The third difference, in some versions, is that the disc does not rotate at a constant speed—instead it slows down as the stylus reaches the outside edge. The reason for this is that it allows each revolution of the disc to give a fixed number of television pictures frames or fields (see page 11). By stopping the stylus from moving in and out, therefore, very good still pictures can be achieved. The problem is that this wastes space on the outer tracks, so playing time is not as long as it might be: systems that rotate the record at a fixed speed give longer playing time but lose this *freeze-frame* facility.

Other disc systems

At the other end of the scale, the simplest system is the RCA Selectavision, a sort of cross between audio and video technology. The disc has a groove in it, like a conventional audio groove, but very many times finer. Because there is a physical groove for the stylus to sit in, no complex control and tracking systems have to be employed. Unlike an audio disc, though, the information is not coded onto the groove walls in the form of physical 'squiggles' (see page 70). Instead, a pit system is again used. The RCA disc is pressed from a conductive plastic material, and the presence and size of the pits is read by a stylus incorporating an electrode which senses the changes in the electrical capacitance of the disc's surface, due to the differing pits.

The RCA system has the benefits of simplicity, but picture quality is not the highest, and freeze-frame facilities are not available. The discs cannot be handled—this would destroy the capacitance effects—and so have to be kept at all times in a protective sleeve, which the video player removes for you as you push it into the machine. Wear might also be a problem.

Falling in between the RCA and Philips systems is another capacitative disc—the VHD. The pits are tracked by electrode as with the RCA system, but there is no groove; instead, there are tracking pits on either side of the signal pits and an error-correction system, of the type used on the Laservision, keeps the electrode stylus on the right path. Freeze-frame and similar facilities are available.

Some video disc player systems play smooth surfaced discs (right), in which the television signal is encoded in the form of microscopic pits (far right), which can be read by a laser.

MEDICAL SCIENCE

BODY SCANNERS

For more than 50 years the simple X-ray photograph was the radiographer's only tool for seeing inside the human body, but in the last decade new systems have been developed. capable of picturing soft tissues as well as bone in the minutest detail. These 'body scanners', as they are known, make it possible to diagnose illness and plan surgery or treatment with greater precision and more quickly and cheaply than ever before.

Body scanners made their debut in 1972. The X-ray *computerized tomographic* (CT) scanner was produced by physicists at the X-ray unit of EMI (Electrical and Musical Industries Ltd) and constituted a major technological breakthrough. A CT machine uses X-rays, but in a sophisticated manner. One of the problems with ordinary X-ray machines is that the pictures they give show only the densest body material, such as bone, with any clarity. Less dense material—muscles, tissue and tumours and other lesions—show up only vaguely, because they absorb very little of the X-rays passing through them. But, although the human eye finds it difficult to distinguish between these areas of little absorption, a computer can be programmed to do the job much better.

The computer in a CT machine is fed data detailing the amounts of X-radiation absorbed by all the types of body material, including water, fat and bone. As a person is scanned by an X-ray beam, it compares this data with the amount of radiation actually absorbed by the material in the person being scanned. In this way, the computer can build up a graphical representation showing the different types of body material in far greater detail than a simple X-ray photograph can.

Rather than exposing a large area of the body to X-rays, as in a normal machine, a body scanner looks at only a thin cross-section or 'slice' of the patient at a time. This also helps to produce a clearer image of the patient's insides.

In the CT scanner, the X-rays passing through the body are picked up by crystal or gas detectors linked to a computer which converts the absorption data into a map of the piece of tissue, visible on a TV screen. The doctor builds up a detailed picture of the area with different cross-sections, which give almost as much information as if the patient were actually sliced up into many pieces. Today's scanners

are capable of examining sections of the body from 2 to 13mm ($\frac{1}{12}$ to $\frac{1}{2}$in) thick, as easily as if they were slices of bread. Seven or eight different slices might be scanned to examine, say, a liver.

No one is quite sure how much damage is done by passing X-rays through the body in the amounts needed for CT scanning. Whereas a traditional X-ray of the head might require 1/10th of a rad (a rad is a unit of absorbed radiation), a similar CT scan would subject the patient to about 4 rads. A scan of another part of the body requires about 1 rad. However, while the CT scanner X-rays a different slice of tissue each time, so that each piece receives only one dose of rays, traditional X-rays overlap so that a patient may receive six or seven times the basic dose in one area of tissue.

Brain examination

The first CT scanners were used for examining the brain. As well as being much more effective in locating an area producing abnormal absorption values (a tumour, for example), than traditional X-ray methods, the scanners also did away with the need for so-called 'invasive techniques' to show up blood vessels and tumours on the brain. These techniques depended on the injection of dyes or radioactive ma-

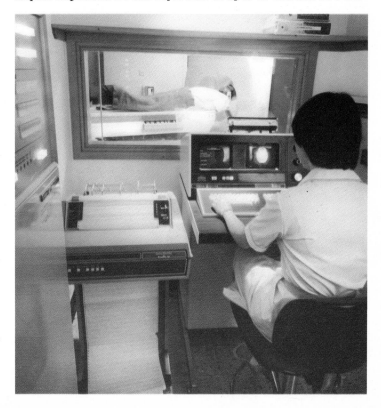

A patient undergoing a CT scan. The computer analysis is shown as a 'density map' on the screen in the middle of the operator's console.

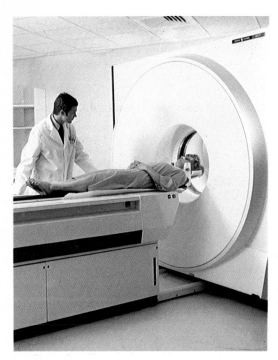

Above and right: To receive a CT scan, a patient must lie on a motorized gantry that locates her precisely within the ring of the scanner. For each section, the X-ray source is activated and sensitive detectors rotate around her body, taking 1½ million readings.

360° BODY SCANNER

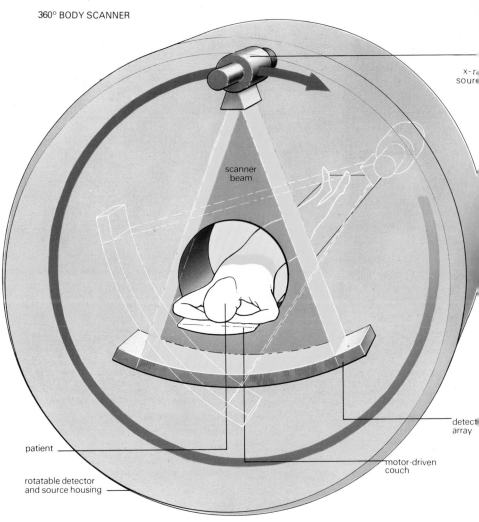

scanner beam

X-ray source

detector array

patient

motor-driven couch

rotatable detector and source housing

terial into the blood going to the brain. Not only was this procedure expensive in time and highly skilled personnel, but it was also uncomfortable and even dangerous for the patient. There is always a chance of doing more harm than good by injecting material into areas as sensitive as the brain.

The system of scanning was later modified so that any area of the body could be placed between the X-ray source and detector. The first scanners could only take pictures of patients lying horizontally, but now there are 'tip-up' and other manoeuvres which enable a patient to be scanned from a variety of angles. This is especially useful for scanning organs which do not lie flat but at an angle, such as the pancreas.

Scanners are improving in many other ways, too. It usually takes about 20 seconds for a scanner to take a look at each slice of tissue, though the latest machines can reduce this to a mere three seconds. The pictures obtained are a lot more detailed than was possible with earlier scanners because a much larger number of absorption points are now plotted. Only a few years ago scanner pictures were made up of only 80,000 points, but most scanners now produce pictures made up of almost four times that number. One model uses 1½ million points for each picture to record extra detail.

One modification of the CT scanner, still being developed, may enable doctors to produce a three-dimensional picture of an area of tissue, so that tumours or blood clots show up 'suspended' in their true positions. Although scanners can look at tissues from a variety of angles they can look from only one direction at a time. What physicists are now trying to do is to take pictures of adjacent slices of tissue and manipulate them by computer so that they can be viewed from all angles, back and front, and as a 3D image projected on the viewing screen.

The major drawback of CT scanners is their enormous cost. Prices range from £250,000 (about $500,000), for the most basic model, to around £1 million for the deluxe version with all the optional but often essential extras.

Sound waves

CT is not the only method for scanning density variations. Another technique is to use very high pitch sound—*ultrasound*. Because the waves used in ultrasound are thought to be completely harmless (unlike X-rays which create a slight radiation hazard), the technique has been used most widely in the field of obstetrics, to locate the developing foetus and to check that there are no limb or other visible malformations, and also to ensure that the baby is growing normally.

As with CT scanning, ultrasound depends on the stimulation of a crystal, this time with an electrical impulse. Each crystal produces sound waves with a specific frequency, which can be detected by another crystal. Ultrasound passes through the tissues and, just as sound waves are reflected as an echo from a cave or mountain as the sound passes from one medium to another, so sound waves are reflected as they pass through tissues of varying densities. These reflections can then be converted into a picture showing up the different densities in an area, so that the womb and the foetus, surrounded by its watery bag of amniotic fluid, can be distinguished from other structures in the same area.

The reflections appear as bright dots of varying intensity on the ultrasound screen. The equipment can be moved around so that the foetus can be seen from all angles, to give a complete picture.

An additional refinement to the system was the introduction of 'real-time' ultrasound in the late 1970s. The part of the machine sending the ultrasound through the tissues, the *transducer*, can be made up of many small parts, each triggered consecutively to give an almost continuous transmission of ultrasound through the tissues. In this way a continuous picture is obtained, much like a ciné film, of the baby as it moves around in the womb.

Ultrasound has also been widely used as a complementary technique to CT scanning to detect tissue lesions. And real-time ultrasound has been proved useful for watching the movement of the heart valves to detect disorders both of the valves themselves and of the heart muscle. It has also been used in liver scanning where tumours as small as a centimetre or two across can be detected. Malignancies of the kidney and pancreas have also been detected by ultrasound, as have abnormalities within the eye. One drawback of the technique is that ultrasound is highly absorbed in air, so any tissues containing air—the lungs or ears for instance—cannot be examined using the technique.

Magnetic scanning

An even more sophisticated (and completely safe) technique called *nuclear magnetic resonance* imaging (NMR) has been developed, and physicists are still investigating its pull potential for revealing diseased tissue.

To understand how NMR works it is necessary to consider what goes on in the most basic component of all material, living and inert—the atom. One way to think of an atom is to imagine it as a planet with moons orbiting around it. The central planet is the *nucleus*, made up of positive particles called *protons* and uncharged particles called *neutrons*. Circling round the nucleus are the *electrons*, the negatively charged particles that balance the charge in the nucleus.

When a piece of tissue is placed between magnets, and radio waves are passed through the tissue, the nuclei of atoms within the tissue will react with the radio waves. Protons absorb and react to these radio waves differently according to their environment, and so the signal which is received after the radio waves have interacted with the tissue can be converted into a picture of the proton density within specific areas of the body.

American researchers were beginning to investigate the potential of the technique on isolated cells at about the same time that EMI scientists were perfecting the first CT scanners. The protons most commonly used are those within hydrogen atoms because these give the strongest signal and hydrogen atoms are found throughout human tissues as components of water. Other atomic absorption patterns being investigated are those of phosphorus and carbon. As researchers master the techniques of the system and work out the optimal magnetic fields and radio-wave frequencies, the pictures produced are becoming much more clear and accurate.

Although there will be some overlap in function between the three techniques of CT scanning, ultrasound and NMR imaging each has its advantages

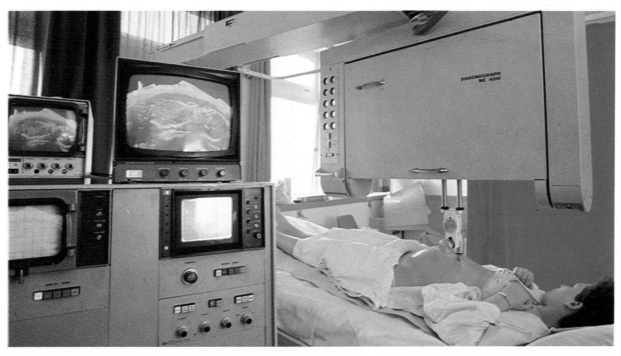

Many women are now undergoing ultrasound scans during pregnancy to detect any foetal abnormalities early. In this picture you can see the outline of womb and foetus on the screen.

ENDOSCOPE

One of the most remarkable tools at the disposal of the modern doctor is an optical instrument which allows him to look right inside the patient's body. He can see into the intestine, stomach, lungs, bladder, womb or even the knee joint. The precision instrument which makes this possible is called an endoscope.

For over 100 years scientists have been developing special instruments for looking at the interior of living organs. The first endoscopes were open, rigid tubes which could be illuminated by candles. Although a few rigid metal endoscopes are still used today, the majority of modern instruments are

Above: An endoscope is used both to view a patient's colon, and to remove a polyp (a small growth) from it. The polyp is removed by tiny forceps at the tip of the instrument, operated by the white handles seen in the picture. A general anaesthetic is not needed, and the procedure takes only a few minutes.

flexible 'fibre optic' types. The key component of a fibre optic instrument is a bundle composed of thousands of long, thin glass fibres. The fibres, 9 to 12 microns in diameter (a micron is a millionth of a metre) are covered with a reflective coating so that however much they are bent or curved, very little light is lost as it travels from one end to the other. The effect of this is that it becomes possible to 'see' round corners: no matter how much the bundle is bent, an image at one end can be viewed by a person at the other.

Fibre optic bundles

A modern endoscope has at least two fibre optic bundles. The first one transmits light from the operator's end of the instrument to the end which is inserted into the patient, the *distal* end. This lights up the area to be inspected. A second fibre optic bundle then transmits an image of the illuminated area back up the endoscope shaft to an eyepiece.

The organization of fibres in the second image bundle is crucial. Because each fibre is transmitting the light responsible for a small bit of the image (rather like a dot on a printed picture) it is very important that the arrangement of fibres is the same at each end of the bundle. A fibre which starts, say, at the top left of one end of the fibre bundle must end at the top left of the other end, otherwise the image seen in the eyepiece will be hopelessly jumbled up. A fibre bundle whose filaments are identically arranged at each end is called a *coherent* bundle.

In most fibre optic endoscopes light is transmitted through the illuminating bundle (which does not need to be of the coherent type) from a powerful lamp, usually a quartz-halogen or xenon arc lamp, to the operator's end of the instrument. From there a second non-coherent bundle transmits it down the flexible shaft to the distal end. A coherent bundle carries the image back up the shaft to the eyepiece. Lenses at each end of the coherent fibre bundle bring the image to a focus in the eyepiece.

As well as the glass fibre bundles, several channels pass through the shaft of the endoscope. One of these allows air to be passed down the shaft. If the organ being examined is a hollow one, such as the stomach, the air channel allows it to be inflated slightly to make thorough examination possible. And as the organ often contains liquid, a suction channel is provided so that it can be drained through the endoscope. Another duct carries fluid for washing mucus or small particles from the viewing lens at the tip.

Control wires are also necessary so that the doctor can point the tip of the endoscope in the direction he wants to view. The tip of a modern endoscope is remarkably flexible and can typically be moved through 100 degrees to the left or right, 180 degrees upwards and 90 degrees downwards. All the controls for these various manoeuvres are conveniently positioned near the eyepiece within easy reach of the operator's fingers. All the channels are completely contained within the shaft, and in spite of the complexity, the shaft diameter may be as small as 8.8mm (0.35in), little more than the thickness of a pencil.

Left: Inside a knee joint; the crossed ligaments are easily seen.

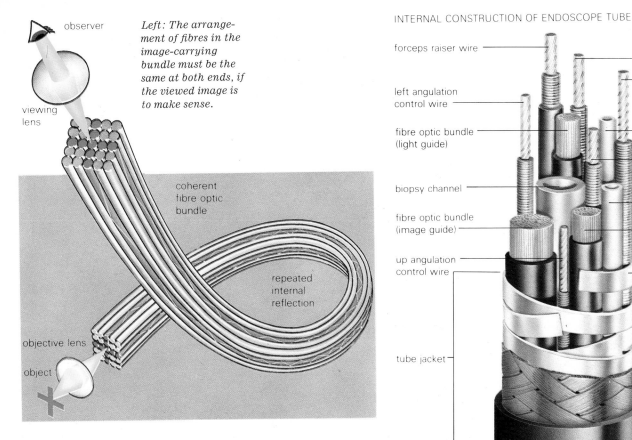

observer

viewing lens

Left: The arrangement of fibres in the image-carrying bundle must be the same at both ends, if the viewed image is to make sense.

coherent fibre optic bundle

repeated internal reflection

objective lens

object

Right: The shaft of an endoscope contains not only fibre optic bundles, but wires to move the tip, pipes to carry air and water, and more.

INTERNAL CONSTRUCTION OF ENDOSCOPE TUBE

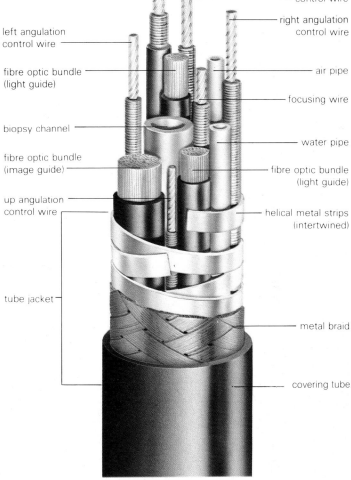

forceps raiser wire

down angulation control wire

right angulation control wire

left angulation control wire

fibre optic bundle (light guide)

air pipe

focusing wire

biopsy channel

water pipe

fibre optic bundle (image guide)

fibre optic bundle (light guide)

up angulation control wire

helical metal strips (intertwined)

tube jacket

metal braid

covering tube

Images seen down the endoscope can be photographed with conventional Polaroid or movie cameras. And more recently a television system has been perfected, so the possibilities for display are virtually limitless.

Specialized endoscopes

The detailed design of the endoscope depends on its intended use. For stomach examination, for instance, the shaft must be long and flexible so that it can easily pass down the patient's throat into the stomach, whereas for knee joint examination the shaft is much shorter and usually rigid. Endoscopes designed for specialized tasks are named accordingly—instruments for looking into the stomach are called *gastroscopes*; for looking into joints *arthroscopes* and so on.

By far the commonest subject for endoscopic examination is the gastro-intestinal tract, the tube which runs from stomach to anus and through which our food passes as it is being digested and absorbed. To view the gullet or stomach, the endoscope is inserted through the mouth, and the patient must fast from the night before the examination. If the large bowel, the *colon*, is to be examined, the instrument (a *colonoscope*) is inserted via the rectum, excreta being removed beforehand by means of enemas and washouts. The doctor inserts the endoscope under local anaesthetic with minimal discomfort to the patient, and after a full examination, which is complete in a few minutes, he has an excellent picture of just what is going on in the suspect organ.

Diagnostic tool

The endoscope is used most often as a diagnostic tool. For instance, if X-ray pictures are difficult to interpret or show nothing unusual, the doctor may turn to the endoscope for a closer look. And if a tumour is revealed by an X-ray picture, the endoscope may be used to take a sample of tissue for analysis before the doctor decides on any treatment. Now that very slender endoscopes are available, some clinics use them as the main diagnostic tool for patients whose symptoms suggest disease of the upper gastro-intestinal tract. And where patients have had an operation on the stomach or colon and their symptoms recur, or where the operation has been for cancer and a regular watch is essential, endoscopy is normally the method chosen.

Endoscopy is sometimes used to examine apparently healthy patients. In Japan, for instance, the incidence of stomach cancer is higher than in other countries, and regular endoscopic examination helps to detect the disease in its

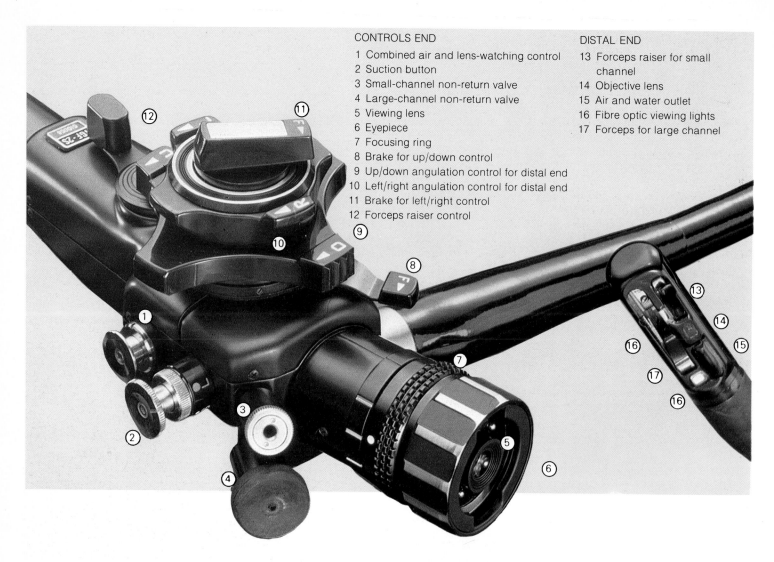

CONTROLS END
1 Combined air and lens-watching control
2 Suction button
3 Small-channel non-return valve
4 Large-channel non-return valve
5 Viewing lens
6 Eyepiece
7 Focusing ring
8 Brake for up/down control
9 Up/down angulation control for distal end
10 Left/right angulation control for distal end
11 Brake for left/right control
12 Forceps raiser control

DISTAL END
13 Forceps raiser for small channel
14 Objective lens
15 Air and water outlet
16 Fibre optic viewing lights
17 Forceps for large channel

early stages and so make a cure more likely. But modern endoscopes are much more than just viewing instruments. Tiny forceps can be passed through yet another channel in the instrument's shaft to take specimens of tissue for analysis. And simple operations can be performed. For example, the traditional method of treating a patient found to have a polyp in the colon involved an operation and a 10 to 14 day stay in hospital. Nowadays the polyp can be removed by means of an endoscope. With the instrument suitably insulated and the patient earthed, a wire is passed down the shaft. A high frequency current is then applied to get rid of the offending polyp. The patient can leave hospital the next day and the procedure is both cheap and safe. A similar wire may be used to enlarge the opening of the bile duct and so allow gall stones contained in it to be removed with a small basket or probe passed down the endoscope. Alternatively, the gall stones may now pass into the bowel of their own accord; either way an operation is avoided. The technique is especially useful for elderly patients unfit for conventional surgery.

All kinds of ingenious modifications of the endoscope and tools for use with it are available so that swallowed foreign bodies may be removed from the gullet and stomach. In the past, if the doctor decided that a particular object would not pass naturally through the gut, he would have no choice but to proceed with an operation. Constrictions of the gullet can be widened by using an endoscope, and it is possible to pass a tube through an inoperable cancer of the gullet to allow the patient to swallow normally despite the disease.

One of the latest developments in endoscopy is a new treatment for patients suffering from a major haemorrhage in the upper gastro-intestinal tract—all too often a cause of death among the elderly. The technique uses a special endoscope fitted with a quartz fibre extending down one of the channels in the shaft. At the operator's end of the instrument the fibre is coupled to a laser light source. First of all the doctor pinpoints the site of the bleeding using the endoscope in the normal way. Then he positions the tip of the quartz fibre over it and triggers the laser for a carefully determined time. The laser light, directed precisely on to the point of the bleeding by the quartz fibre, coagulates the blood and the bleeding stops.

Developments like this will ensure that the endoscope remains a powerful weapon in the doctor's armoury. And in the future it can be expected that cheap, fast endoscopic treatment will increasingly replace conventional surgery.

INTERFERON

Research into a mysterious substance called interferon is one of the most active lines of medical investigation in the 1980s. Interferon, the generic name for a group of proteins involved in the body's defence against disease, is released naturally by cells in response to attack by viruses. The proteins seem to prevent further infection—probably acting similarly to antibiotics. But it is the suggestion that interferon may also be valuable in the treatment of cancer that has caused the excitement.

The interferon story began in 1957, when Alick Isaacs and Jean Lindemann, working at the National Institute for Medical Research in England, isolated a substance which seemed to prevent virus infections. They had artificially infected cells in the membrane of a chicken's egg with a certain type of virus, and were trying to discover why it was impossible to infect the same cells with other kinds of viruses introduced afterwards.

They found that an extract taken from the infected cells, and added to a separate group of completely healthy cells, caused the second group to become resistant to a wide range of virus infections. It was soon shown that the active ingredient in the extract was a protein; and because it seemed to interfere with a virus's ability to infect cells, Isaacs called the protein *interferon*.

Viruses work by invading a cell, taking over genetic control of its biochemical systems, and using those systems to produce identical copies of themselves. The normal functioning of the infected cell is so disturbed that it rapidly disintegrates, releasing all these newly manufactured viruses which can infect other cells. However, one of the last actions of the dying cell is to release a minute quantity of interferon into the surrounding fluid and bloodstream. Molecules of interferon then attach themselves to special receptor sites on the surfaces of healthy cells, causing the production of

Right: This researcher is inserting the gene for human interferon into bacteria. The hope is that, by this genetic engineering, the bacteria will produce human interferon as they grow and divide—yielding the large amounts of the protein that are necessary if properly conducted trials of the drug's effect are to be staged.

Below: In close-up, the needle penetrates a frog's egg to inject RNA in a test to see if the genetic engineering worked.

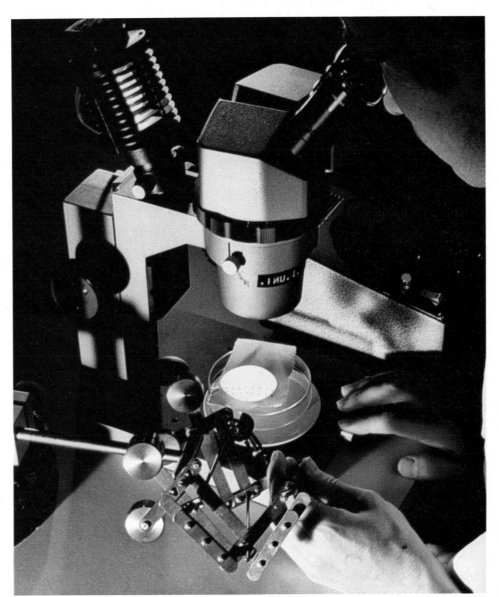

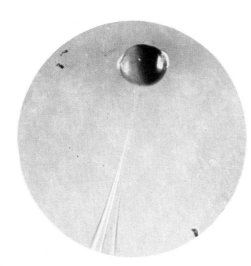

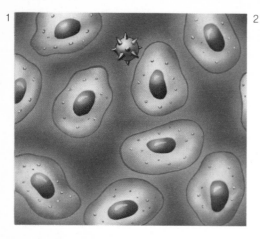
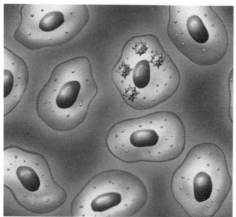
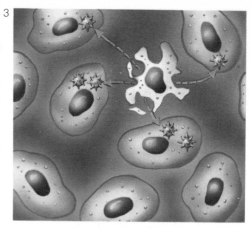

enzymes which in turn help protect the cells from virus attack.

From a clinical point of view, it is significant that once the body has been 'alerted' by interferon, its cells become resistant to a wide variety of viruses—not just to the particular infection which sparked off the reaction. The earliest samples of human interferon studied confirmed that the substance could indeed help control viral infections. A single nasal spray containing interferon gave protection for up to two weeks against the common flu virus, and doctors at Moorfields Eye Hospital, England showed that it could protect victims of shingles against blindness.

The discovery of a substance that was potentially a cure for all virus-caused diseases was greeted with excitement, and many laboratories began experimenting with interferon. However, it soon became clear that there were obstacles to its immediate use.

For a start, scientists found that the precise chemical structure of the substance varied critically from species to species; as a result, interferon from animals simply would not work in humans. Obtaining human interferon proved a daunting task, partly because interferon is present only in extremely small concentrations, and partly because it is far from easy to purify.

Fighting cancer

Hoping to demonstrate that interferon would also work against a cancer-causing virus in mice, American scientist Ion Gressor tried injecting the mice with mouse interferon. He expected the interferon to prevent the further spread of the tumours, but to his astonishment, the tumours actually shrank—and in some cases disappeared altogether.

Funds for further research became rapidly available. Soon it was shown that in animals, and in large doses, interferon appeared to have a two-fold effect against cancer. In the first place it activated the scavenging white cells in the blood stream which normally destroy foreign material in the body. Second, the surface coating of tumour cells which have been exposed to interferon undergo a slight change. The effect of the change is that cancer cells become more recognizable to the body's natural immune system as foreign material, and this encourages their destruction.

The first systematic experiments took place at the Karolinska Hospital in Sweden during the mid-1970s. There, 35 patients with a rare form of bone cancer were treated three times a week for 18 months. More than twice the expected number of patients showed long-term recovery—but the number of people treated was too small to draw any general conclusions about the real worth of the treatment. Many doctors were sceptical, and saw interferon as just another in along line of 'wonder drugs' that failed to live up to their early expectations.

Inconclusive trials

So the American Cancer Society funded a much more detailed program—involving 150 patients, costing millions of dollars, and using up almost the whole world supply of interferon. Yet even these trials have proved inconclusive. Some cancers improved; others did not.

It is clear that the main restraint on interferon research is the lack of an adequate method of production. As a result, hope for the future lies in finding a large-scale, cheaper source of the drug. The aim is to produce interferon artificially in quantities which are not restricted by the need for human raw materials. So far three different approaches to the problem look promising.

The method which is furthest advanced is to culture human white blood cells that have lost their normal growth control and multiply readily. One problem is that the best way of making the blood cells multiply is to expose them to a special virus. This virus has been implicated in certain types of human cancer, so it is vital that the interferon produced is uncontaminated with an agent that will cause the very disease that the drug is designed to cure. There will also be difficulties in scaling-up production to large-scale commercial cultures.

A variation of this approach is to dispense with white cells and use fibroblast cells instead. These cells, which can easily be obtained from human foreskins removed during circumcision, will multiply in culture without being treated with dangerous viruses. While the interferon produced by this method is slightly different from white blood cell interferon, it seems to work just as well. Indeed, some researchers claim that a mixture of types is most effective.

Genetic engineering

A quite different artificial method of producing interferon in large quantities has become possible as a result of recent

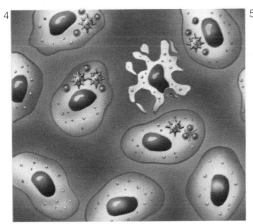
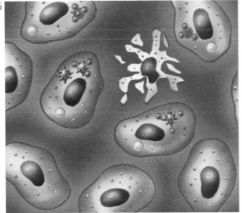
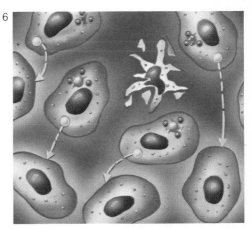

advances in genetic engineering. A section of DNA, representing the gene coding for human interferon, has been successfully inserted into the DNA of a bacterium, *Escherichia coli*, which then produces interferon as it grows and divides. The major problems concern the extraction and purification of the interferon so produced, but these problems seem likely to be overcome.

The most direct solution to the problem of interferon supply would be the achievement of complete chemical synthesis. Before this is possible, the precise sequence of 150 or so amino acids that comprise the interferon protein must be known. This, in turn, depends on having extremely pure samples of human interferon, and material of this quality has only recently become available.

Both the direct synthetic production of interferon and the use of bacterial 'chemical factories' to achieve the same result may well come up against a serious obstacle. Interferon is not simply a protein—it also contains carbohydrate components known as sugar residues. The exact nature of these residues is not determined by the gene for interferon. Instead, the chemical constitution of the interferon molecule is partly affected by the type of cell that makes it.

It is possible that interferon from bacteria which have undergone genetic engineering and interferon from straightforward chemical synthesis will both be critically different from natural human interferon. Whether this difference significantly reduces the power of artificial interferon to help destroy cancer is unknown. It may be responsible for the inability of interferon from one species to act properly in the body of another.

Interferon's future

Interferon is unlikely to be a cure-all for cancer, despite the hopes raised by some television and press reports—some researchers, too. But it is vital that research continues so that the real potential of this medical tool is properly exploited.

For example, less exotic examples of dangerous diseases which may be cured by interferon include hepatitis B and acute chicken-pox in children. The drug may also play a useful role in the aftermath of organ transplants, when administered in dosages which boost the body's defences against infection without stimulating the immune mechanisms that lead to tissue rejection.

Above and right: This sequence of diagrams shows the natural process by which interferon protects cells against attack by viruses.
1 A virus prepares to enter a cell.
2 The virus enters the cell and divides.
3 The cell breaks up, releasing the new virus particles which infect other cells.
4 Infected cells respond by manufacturing anti-viral protein to attack the viruses.
5 Infected cells also produce interferon.
6 The interferon is released to surround other, unprotected, cells.
7 These then react to the interferon by producing anti-viral protein, and thus become resistant to attack by viruses.

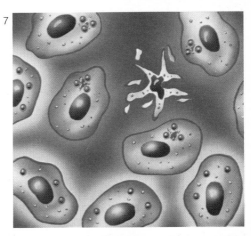

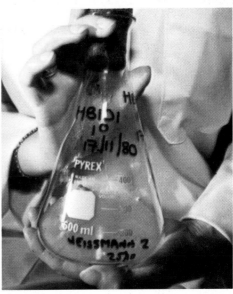

Above: At Imperial College, London, England, this flask contains bacteria genetically engineered to produce human interferon.

KIDNEY MACHINE

Failure of a person's kidneys is fatal. The safest solution is a kidney transplant—an operation with a high rate of success. But not every kidney patient is a suitable candidate and, unfortunately, there is not an unlimited supply of suitable donor kidneys. For many patients, the life-saving alternative is a mechanical kidney machine.

The function of the kidneys is to filter the blood and remove impurities, which would become poisonous if allowed to accumulate. The kidneys do this in two stages, first using a coarse filtering mechanism called the glomerulus, and secondly a fine adjustment, which takes place in the tubules, where the important nutrients which escape into the collecting system are absorbed back again. The kidney machine is a copy of the glomerulus, but it cannot carry out the function of the tubules.

Cleaning the blood

The basic principle on which kidney machines work is very simple. A very thin membrane, or 'sieve', is placed between the blood and cleaning fluid, which is called *dialysis fluid*. Blood is very high in its concentration of certain impurities, whereas the dialysis fluid is low. This results in a 'concentration gradient', or natural tendency for the impurities to flow across the membrane into the dialysis fluid until the concentrations on each side of it are balanced, rather like the weights on a pair of scales. The size of the holes in the membrane determine the size of the molecules of the impurity which may pass across. The smallest molecules pass most readily and large molecules, such as albumin protein cannot pass the membrane at all.

In the past, flat sheets of cellophane were used in kidney machines, as the dialyzing membrane and blood was pumped over one side while the other side was bathed in dialyzing fluid. These Kil plates, as they were called after the Scandinavian doctor who invented them, used up a lot of space and more compact arrangements, using coils of membrane, are now built into kidney machines; this makes the machines smaller so patients can now undergo dialysis in their own homes. Dialysis membranes used to have to be cleaned repeatedly, but disposable units are now available, greatly cutting down the risk of infection.

Not only the size of molecules, but the speed at which they pass is important. Molecules which have a very low molecular weight have a high rate of clearance from the blood, whereas the ones with a high molecular weight have a low rate of clearance. What is needed, therefore, is the largest possible area of membrane together with a good supply of blood to run across it.

To ensure an adequate and ready supply of blood for cleaning, this must be taken from a large blood vessel—usually an artery in the forearm is used for this purpose. There are two ways this can be done. The first, which is a temporary arrangement, involves inserting a little plastic tube into an artery and another into a vein in the forearm and connecting these by means of more tubing to the kidney machine. This is known as a shunt, and when it is not being used for dialysis,

the tube in the artery is connected to the tube in the vein, so that the blood continues to flow and does not clot.

A more permanent arrangement is known as a *fistula*. Here, a small operation is performed to join an artery to a vein under the skin surface. After about six weeks the vein becomes very thick-walled and can then be punctured repeatedly with needles, which are connected, again with plastic tubing, to the machine. The quality and maintenance of the shunt or fistula is essential to the patient's well-being.

To stop the blood clotting in the kidney machine, the anti-clotting substance heparin is used. It is very important that no air bubbles get into the blood and pass back into the patient, because these may cause blockage of small vessels in the lungs. A special bubble detector ensures that blood is free of air bubbles. A pump is required to drive blood through the membrane, and the necessary fluids must be warmed. Cold blood going back to the patient could quickly cause hypothermia.

Home dialysis

Patients are taught exactly how to look after themselves during home dialysis. For example, the shunt must be kept clear of blood clots by flushing it through with blood thinners from time to time, and the skin over the area needs to be kept scrupulously clean in order to avoid any risk of infection.

The frequency with which a patient needs to be connected to the dialysis machine depends upon the severity of his or her kidney disease (even in severe renal-kidney failure the kidneys are usually capable of filtering some urine), and the quality of the shunt. A shunt which delivers a good volume of blood to the machine will enable the blood to be cleared of impurities more rapidly than a poor shunt. It is not unusual for patients to need two to three periods of dialysis each week, each treatment period lasting up to five hours.

Other problems

All patients with chronic renal failure become anaemic, and dialysis does not correct this. At least one of the causes of this anaemia is a deficiency of a stimulant to blood production in bone marrow; this is a hormone called erythropoietin, which is made in the kidneys. The anaemia can be treated by blood transfusion, but this does not remove the problem and when the transfused cells die off in about 50 to 60 days, the anaemia must be treated all over again.

Raised blood pressure can sometimes be a problem. It may be treated with conventional drugs, but more often the doctor attempts to reduce blood pressure by lowering the amount of body salt. This can be done either by restricting the amount of salt in the diet or putting less salt in the dialysis fluid.

It is known that raised blood pressure causes an increased risk of blood vessel disease and there is a higher incidence of heart attacks and strokes amongst patients on dialysis than in other people.

Risk of infection

Patients with chronic renal failure are more prone to infection than others. Chest and urinary infections cause most trouble. Much more hazardous, particularly for the staff

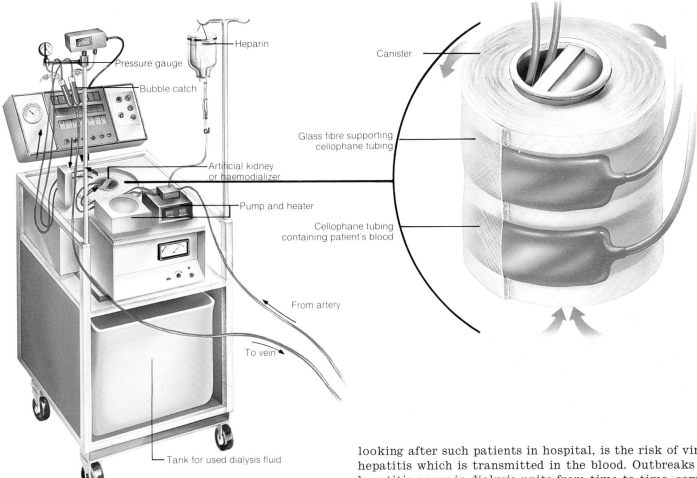

Heparin

Pressure gauge

Bubble catch

Artificial kidney or haemodializer

Pump and heater

From artery

To vein

Tank for used dialysis fluid

Canister

Glass fibre supporting cellophane tubing

Cellophane tubing containing patient's blood

Above: A typical kidney machine. Blood for cleansing flows from patient's artery into the haemodializer; then through two coils of flat cellophane tubing, immersed in a canister through which the dialysis fluid is pumped, so washing out waste products from the blood. Purified blood returns to patient's vein.

Below: A patient with this portable kidney machine can go on holiday, taking the machine along. All that is needed is a hatch-back car and a power source.

looking after such patients in hospital, is the risk of viral hepatitis which is transmitted in the blood. Outbreaks of hepatitis occur in dialysis units from time to time, sometimes with fatal results. For this reason, all blood dialysis fluid is handled with great care. Gloves and masks are worn at all times and spillages of fluid are cleaned up immediately because of the risk of inhaling the hepatitis virus. Unfortunately, because of the risks to staff, some dialysis units may be reluctant to take patients who are hepatitis carriers.

Side effects

Although dialysis is a life-saving procedure, it takes time and may be unpleasant for the patient. Home dialysis has improved the inconvenience of regular admission to hospital, but the stress it places on the patient's family cannot be underestimated. Depression may be a severe problem with patients dependent on dialysis. However, the support and encouragement of a patient's spouse and family can do much to alleviate this, as can professional therapy.

Dialysis is usually a second choice to a kidney transplant and is often used only until a suitable donor kidney becomes available.

Clearly the most encouraging development for those patients awaiting a donor kidney—or who are undecided about having a transplant—is the improvement of dialysis membranes, the technology of which is rapidly advancing. Improved membranes will shorten the time patients have to spend undergoing dialysis and will thus lessen the inconvenience and the depressing effects.

NUCLEAR MEDICINE

There are many ways of making images of the body's interior, including conventional X-rays, computerized scans and the use of high-frequency sound (see pages 77 to 79). But nuclear medicine has one major advantage over the other imaging methods: it can find out how an organ actually functions.

For example, a conventional X-ray may show that a patient's left and right kidneys both *look* normal, but nuclear medicine imaging techniques might show that one of them was, in fact, *performing* very abnormally.

In nuclear medicine, a radioactive source is taken into the body in some way. As the source passes through the body, it emits gamma radiation (short-wavelength X-rays) which can be detected by a gamma camera outside the body. The camera is used to map out the movement of the source and the places where the gamma-emitting radioactive material has been deposited. One major advantage of this technique

is that the mapping is done by watching the *emission* of radioactive material from the body, unlike X-ray work, which relies on the body *absorbing* radioactive material. In nuclear medicine, most of the radiation escapes from the body, so that the dose to the patient is often much lower than with a conventional chest X-ray.

The source can be of various types and may be introduced into the body in a number of ways—for example, by injection —and is called a *radiopharmaceutical*. The radioactive material (called a *radionuclide*) in this pharmaceutical will also vary, depending, for example, on what part of the body is under investigation.

For example, to test whether a kidney is functioning, a pharmaceutical which will be filtered out of the blood is used, coupled to a radionuclide whose radiation is readily detectable. If, and only if, the kidney is working will it filter out the chemical which will then show up as a 'hot spot' on a radionuclide scan.

The most common investigation carried out in a nuclear medicine department is the bone scan. In a normal scan, the manufacture of bone can be seen to be occurring throughout

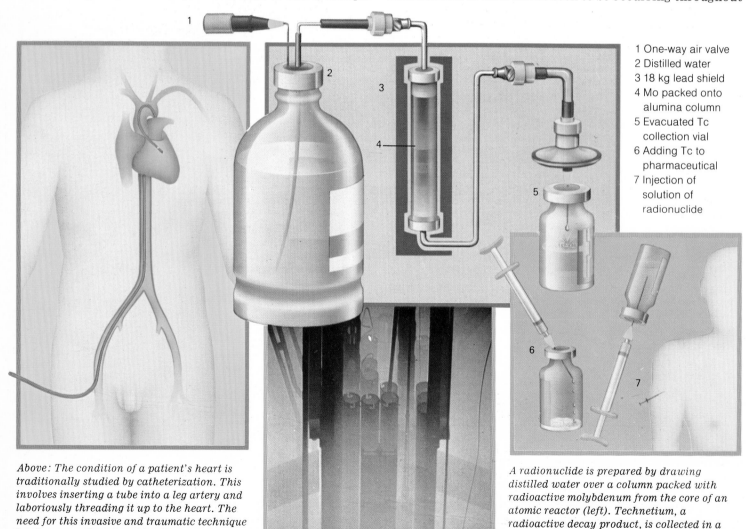

1 One-way air valve
2 Distilled water
3 18 kg lead shield
4 Mo packed onto alumina column
5 Evacuated Tc collection vial
6 Adding Tc to pharmaceutical
7 Injection of solution of radionuclide

Above: The condition of a patient's heart is traditionally studied by catheterization. This involves inserting a tube into a leg artery and laboriously threading it up to the heart. The need for this invasive and traumatic technique has been greatly reduced by the recent advances in the field of nuclear medicine.

A radionuclide is prepared by drawing distilled water over a column packed with radioactive molybdenum from the core of an atomic reactor (left). Technetium, a radioactive decay product, is collected in a bottle, drawn into a syringe, and then injected into a bottle with the pharmaceutical to be labelled. This is then administered by injection.

the skeleton, and particularly in the hips, spine and shoulders. But the bone scan can show up a particular problem such as an ill-fitting artificial hip. The artificial hip causes increased wear on the surrounding bone, and leads to a local increase in the rate of bone remodelling. The increase in remodelling is revealed as a tell-tale 'hot spot' on a nuclear scan since it leads to an increase in the uptake of radiopharmaceutical.

This type of investigation is known as a *static scan*—an image of the radiopharmaceutical distribution is obtained at just one moment in time.

Lung test

Another example of a static scan involves injecting very small spheres of human albumin labelled with technetium. The size of the spheres is chosen so that they will be carried in the bloodstream as far as the capillaries (the smallest-diameter blood vessels) in the lung, where absorption of oxygen takes place. The spheres are, however, too large to pass through the capillaries and lodge there in the lung.

In a normal image, the uniform deposition of radioactive spheres indicates that the blood supply to the lungs is intact. If, however, there is no radioactivity present in, say, the mid-zone of the right lung, the test indicates that there is no blood supply to this region.

Such a clearly defined defect results from a pulmonary embolism (in which a large blood clot blocks one of the arteries supplying blood to the lung). If the patient is treated with anti-coagulants, the clot may be broken up and the blood supply restored. Thus the lung scan can be used, first to help make the diagnosis, and then to monitor treatment.

Dynamic scans

Often, it is important to monitor the effects of the pharmaceutical over a period of time, and this is done by a *dynamic scan*. Images are taken one by one over a period of time, usually starting with the administration of the pharmaceutical, and can then be viewed in the same sequence, rather like a strip of movie film. The shape of the image will indicate where the radiopharmaceutical is at any particular moment, and the brightness of the image indicates the rate at which it is being absorbed by the body. A typical dynamic study is the *renogram*—the investigation of kidney function. The kidney is a collection of about a million tiny filter units, called *nephrons*. These remove certain waste products and foreign materials from the bloodstream, concentrate them, and pass them to the bladder in urine. In order to test the filtering function, a chemical, such as hippuric acid labelled with radioactive iodine, is injected into the blood stream.

A series of images is collected at 20-second intervals starting at the time of injection, and the gamma camera is connected to a computer which stores the images. For each image in the series, the computer calculates the amount of radioactivity present within the kidney and displays the result in the form of curves known as activity-time curves.

The curve rises initially, corresponding to the extraction of the radiopharmaceutical from the blood, before output into the urine takes place. Then the curve reaches its peak (as the rate of extraction is balanced by the rate of outflow into urine). Finally, the curve falls (corresponding to a phase

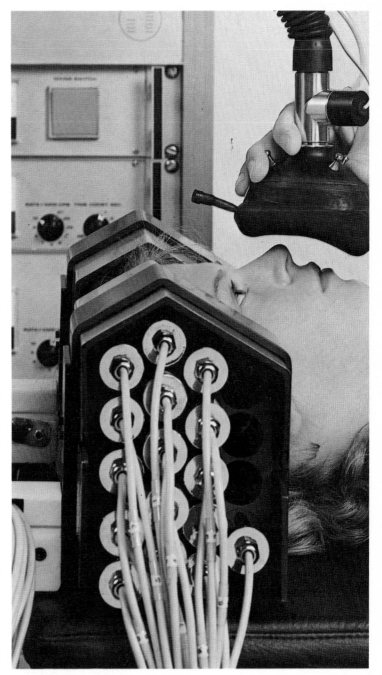

The state of health of the brain is monitored by the inhalation of radioactively labelled xenon gas. Detectors at the side of the head monitor the rate of clearance of the gas—thus measuring the level of blood flow in the brain. Results are displayed on a video monitor (left).

A single crystal of sodium iodide—some 450mm (18in) in diameter and 13mm (½in) thick—is the heart of the gamma camera. It is the point at which radiation is actually detected. Gamma rays, unlike light, cannot be focused by a lens. To select which rays will be used to create the image, unwanted rays are absorbed in a block of lead, called a collimator. Holes drilled through the block (fitted to the patient side of the crystal) allow only certain rays to pass through to form the gamma ray image.

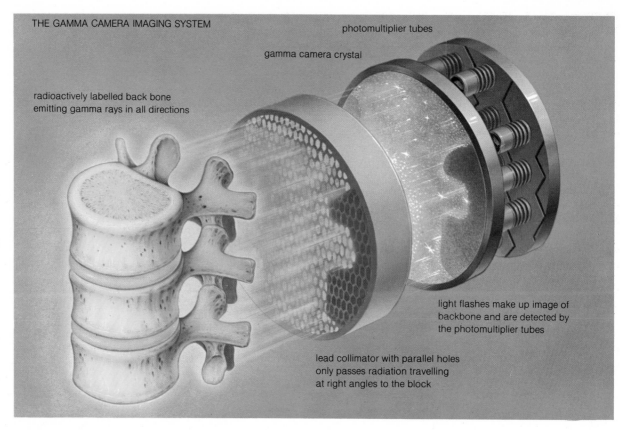

THE GAMMA CAMERA IMAGING SYSTEM

photomultiplier tubes

gamma camera crystal

radioactively labelled back bone emitting gamma rays in all directions

light flashes make up image of backbone and are detected by the photomultiplier tubes

lead collimator with parallel holes only passes radiation travelling at right angles to the block

in which the radiopharmaceutical is excreted from the kidney). Any variation from this pattern—such as when the kidney is not filtering material from the blood, or when the filtering function is intact but the kidney is not excreting the material into the urine—produces a distinctive curve.

Nuclear angiograms

The fastest-growing area in nuclear medicine is the radionuclide investigation of heart function. Nuclear cardiology offers the cardiologist a much less disturbing method of obtaining data than the standard technique, which is called *contrast angiography*. This traditional method involved inserting a hollow tube (catheter) into an artery at the thigh and passing it along until it reached the chambers of the heart. A radio-opaque dye is then injected into the heart via the catheter and a series of X-ray images obtained. This technique of angiography involves a small risk of serious complications, requires hospitalization, and is traumatic.

A nuclear angiogram can be performed by passing a pulse of labelled material through the heart (the 'first-pass' technique). With this method, a gamma camera is set up to record 25–100 images per second of the passage of the radionuclide in the heart.

The gamma camera is positioned so that it records activity in both chambers of the heart. From the gamma camera image it is then possible to calculate the time taken for the blood to travel from one side of the heart to the other. It is also possible to feed into a computer information from successive beats of the heart. The computer can then be instructed to combine the images to produce an accurate representative cardiac cycle, which is displayed in the form

of a moving pattern on a television screen. This pattern of movement can be used, for example, to identify a patient with an *aneurysm*—a segment of heart muscle which has died and hardened following a heart attack. Radionuclide tests cannot always supply all the data that can be obtained by conventional angiography, but they do still have a wide range of applications.

Beta radiation

Nuclear medicine also uses beta radiation, a short-range killer of cells. It comprises a stream of electrons, which are rapidly absorbed in tissue (having a maximum range of a few millimetres). Because the electrons give up their energy over such a short distance, they deliver a high-radiation dose that kills the tissues through which they pass.

But this cell-killing ability can also be exploited for safe and non-traumatic therapy. If the radionuclide is aimed so that its lethal effects are restricted to the site of a tumour, the beta radiation will destroy the cancer cells without damaging the normal healthy surrounding cells. For example, if radioactive iodine is administered, the body will tend to concentrate it in the thyroid gland just as it concentrates normal iodine. Thus the radio-nuclide finds its way to this small target where it kills cancer cells.

Nuclear medicine procedures are not traumatic for the patient, most requiring only a small injection into an arm vein. Virtually all tests can be performed without having to hospitalize the patient, and the procedures have no side-effects. Nuclear medicine offers a safe and convenient method of obtaining reliable information about the function, as well as the structure, of a wide range of body organs.

OPERATING THEATRE

An operating theatre is the most critical environment most people ever experience. It must be designed to enable the surgical team to work with maximum efficiency, exposing the patient to minimum risk—especially from infection to the surgical wound. Design of an operating theatre starts with the design of the hospital itself, and its location is discussed by architects, hospital authorities and surgeons.

An operating theatre must provide a completely *aseptic* (germ-free) area—near to the surgical wards but isolated enough to be undisturbed by the day-to-day work of the hospital. Between the wards and the operating zone there must be a transfer area, enabling movement of patients from the 'unclean' general areas of the hospital to the operating department with no risk of spreading infection. After an operation the patient will go first to the recovery room, then back via the transfer area to the ward. Meanwhile, all unclean material must be hygienically removed from the operating area.

The details of how hospitals achieve these basic aims vary. British surgeons prefer a system that involves three main zones: (1) a transfer area; (2) space for surgical arrangements such as anaesthesia preparation, sterile stores, and scrub room; and (3) the operating theatres themselves.

Preparing for surgery

The surgeons, nursing staff and patients all reach the operating area via an unclean corridor from the main hospital. The surgical and nursing staff enter separate suites of anterooms with lockers, lavatories and rest facilities. There they change from day clothes into surgical gowns, caps, masks, boots and overshoes. They do not shower because showering causes the body to increase the rate at which it sheds contaminated, bacteria-bearing skin particles. From the preparation area, the medical teams go first to the scrub room, to clean their hands and forearms, and then to the operating theatre.

The patient, meanwhile, has entered the operating zone through a transfer room. This critical area must be designed so that disease-causing germs are not carried into the clean zone from the rest of the hospital, either on the trolley wheels, on orderlies' shoes or on the patient himself. So the transfer area has a demarcation line where the 'unclean' and 'clean' areas meet—this can be a simple line on the floor. The patient is transferred from his bed or trolley by the ward staff on the unclean side of the line to an operating table-top on a transport carriage on the clean side.

This system does not guarantee that germ-bearing footwear worn by the ward staff will not stray from the unclean side of the line. A low wall, perhaps two feet high, is more effective, but the most sophisticated systems use a hatchway, built into a stainless-steel partition, to divide the transfer room. Using a hatchway system, orderlies bring the patient on a trolley to the partition. An operating table-top is ready beyond it, on its own carriage. A lifting arrangement raises the table-top, opens the hatchway window and swings the table-top into the unclean area. The patient can either move across to the operating table himself or be carried. The operating table then swivels back into the clean area, and the theatre staff wheel it first to the anaesthesia preparation room and then to the theatre itself. While the hatchway window is open, positive air pressure on the clean side forces a constant current of pure air out through the hatchway, keeping unfiltered air and bacteria on the unclean side.

After the operation, the process is reversed. The patient is returned first to the transfer area and then to the recovery area. Doctors and nurses return to their respective locker rooms. Soiled stores and instruments are taken to a disposal point to be sterilized or discarded.

Fast turnover

The advantage of this system lies in fast turnover; one patient can be undergoing anaesthesia while another is on the operating table. There is also a psychological advantage: some surgeons believe that it would disturb the patient to enter the theatre before being anaesthetized.

American surgery departments employ similar patient transfer systems, but they have no separate anaesthesia preparation room. The patient arrives in the operating theatre and is subjected to the full splendour of the operating paraphernalia—gowns and masked surgeons, plus theatre lights, ancillary equipment and instrument trolley —before being anaesthetized.

A modern operating theatre—clean, sterile and brightly lit—with an operation in progress.

Clean room technology

All surgical operations require pure air to eliminate the risk of airborne infection. Until recently, some operations (such as hip-joint replacement) have been too hazardous to undertake because deep-wound infection could be disastrous. But developments in clean-room technology can now provide a virtually sterile atmosphere.

One basic requirement is to have as few people as possible in the operating room. This can be achieved by using a glass-walled cabin into which only the operating team and the patient are allowed. The ancillary staff, including the anaesthetist, remain outside the cabin with their equipment.

Modern theatres also have pure, filtered air continuously flowing down onto the operating table, then away and back to high-efficiency particle-arrest filters.

One advanced system, developed in the 1960s, abolishes the walls of the operating theatre altogether. Instead, patient and surgical team are enveloped in a column of clean, filtered air which sweeps downwards from a canopy above the operating table, covering over an area of about 9 sq m (30 sq ft). Contaminated air circulates upwards, away from the operating table and back to the filtration plant in the ceiling.

Surgeons and other theatre personnel are also a source of potential infection. The human body, especially in the heat and stress of an operating theatre, emits about a thousand scales of skin per minute—all carrying organisms which could circulate on the convection currents of the operating theatre. A 'total body exhaust' system can eliminate this source of potential infection. The surgeons and nurses do not wear the familiar cap and breathing mask. Instead, they wear an all-enveloping, one-piece hooded gown with a lightweight visor. The gown is made of closely woven material to prevent the passage of particles as small as 4 to 20 microns in size. (Cloth in conventional gowns blocks only particles over 50 microns.) The particles are drawn off through a flexible tube under the visor. This puts a negative pressure under the surgical gown, so a continuous upward flow of purified air bathes the surgical teams' bodies, keeping them fresh and helping to stave off fatigue.

Intensive care rooms

It is equally important to supply pure air—at the right temperature and humidity—in the post-operative intensive-care rooms, where a patient may be vulnerable to infection for several weeks. Burn patients must often lie with their wounds exposed, and cancer and leukaemia patients often receive drugs which reduce their natural resistance to infection. In isolation hospitals, where infectious conditions are treated, it is vital to maintain a supply of purified air to protect everybody in the hospital from being infected by the patient. In such a case, the air-treatment plant keeps the patient's room at negative pressure. Air flows into the room from outside, and air used by the patient is carried away to be filtered.

Operating tables

The surgeon's next need is for efficient equipment. The operating table is a main priority. At its most basic, it could be a simple, cleanable table. It should also be adjustable, so the surgeon can bring the patient's body to the exact position he wants for unimpeded access to the operating site.

Simple and comparatively cheap models will double as operating table and trolley. The patient stays on it from the time he enters the clean zone, through anaesthesia and the operation itself, and during the immediate recovery period.

Virtually all tables will have a facility for raising the patient's head or feet. Many include other devices to support the patient in the precise position required: leg and body supports and arm tables can all be attached to the sides of the main table if needed.

Inside the table itself, a compartment will carry X-ray cassettes and an image intensification device. Part of the table will be X-ray translucent so that the surgeon can have up-to-date X-ray information during the course of the operation.

A recently developed operating table hinges at four points to divide into five supporting panels. Electrically controlled hydraulics adjust the table at points corresponding with the four main articulation points of the human body: the neck, lumbar arch, hip and knee. The surgeon can adjust the table to position the patient exactly as required. Controls include a lateral tilt to each side and a facility to raise and lower the entire table.

The most sophisticated operating tables also include a detachable top, which fits over a fixed, centrally placed supporting pillar. Piped oxygen, compressed air, medical vacuum and steam can be provided through this pillar.

Surgical light

Finally, the operating theatre must be well lit. The main surgical light, about 20 times as much as falls on an average office desk, is provided from a bank of moveable lamps over the operating table. Light is directed onto the patient from

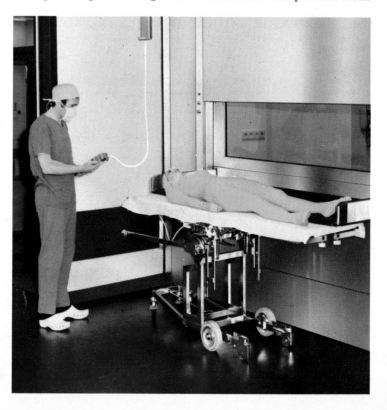

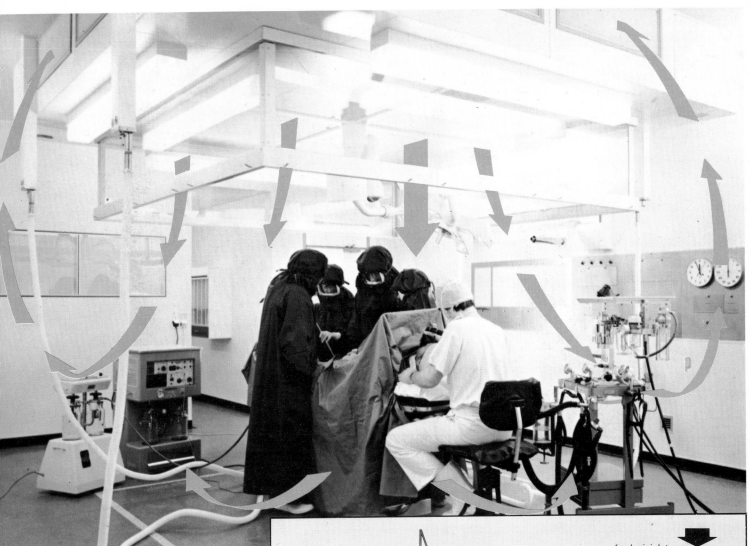

Above: The clean air system surrounds the surgical team in a column of filtered, sterile air which sweeps down continuously from an overhead system of fans and filters. Contaminated air circulates upwards and back to the ceiling unit. The surgical team wear one-piece hooded gowns fitted with plastic visors. Flexible tubes under the visors connected to suction points remove bacteria and bathe the wearers' bodies in a clean air flow. An audio system ensures clear communications. The anaesthetist is outside the clean air zone, so he can wear a conventional gown and cap.

Right: An alternative clean air system. The octagonal shape encourages a quick, even flow of sterile air. Here, the surgeons wear conventional caps, gowns and masks.

Left: A transfer hatch ensures sterile conditions. And the motor-driven rollers move the patient easily—reducing effort for hospital staff.

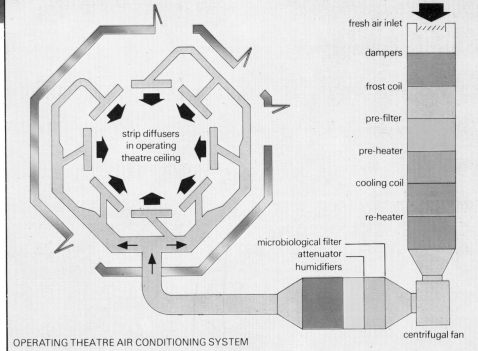

fresh air inlet

dampers

frost coil

pre-filter

pre-heater

cooling coil

re-heater

strip diffusers in operating theatre ceiling

microbiological filter
attenuator
humidifiers

centrifugal fan

OPERATING THEATRE AIR CONDITIONING SYSTEM

several sources so that plenty of strong light passes round the surgeon's head and shoulders. In this way, the surgeon avoids throwing his own shadow on the surgical wound when he bends over the patient.

Surgical lamps are arranged to produce the right pattern of light, concentrating on an area about 250mm (10in) in diameter, with reasonable brightness over a range of between 750mm and 1500mm from the lamps. At one time, it was necessary to use a transformer to achieve the intensity of light needed for surgery. Now, quartz-halogen lights allow the surgeon the right degree of intensity—ten times that of a domestic incandescent bulb—from the normal voltage. Modern lighting also employs solid state intensity control, so that the surgeon can adjust light intensity.

The intensity of light is not the surgeons' only consideration. Surgeons can suffer serious eye fatigue in the course of a long operation, especially if their eyes are forced to move between highly contrasting areas—from shiny steel instruments, white swabs and white operating theatre walls to the darker tones of the surgical cavity. The difficulty can be alleviated by painting the walls in a neutral glare-free colour, using dark surgical drapes and manufacturing surgical instruments with a satin finish to reduce their light-reflecting properties.

Accurate *colour rendering* is also crucial: the lighting should produce results as near to daylight as possible. Ordinary light bulbs give a yellowish light, lowering the eye's ability to distinguish subtle colour differences in the blue range. Modern surgical lights have a colour rendering about as close to daylight as artificial light may get. For some

time, theatres were built without windows, but the trend today is to allow in some daylight if the building arrangement makes that possible. Surgical teams tire less quickly and suffer less eyestrain under natural light, and of course it helps colour rendering.

Within an accurately controlled and planned environment, the surgical team of a modern hospital enjoys a high level of freedom to concentrate on exercising their life-saving skills.

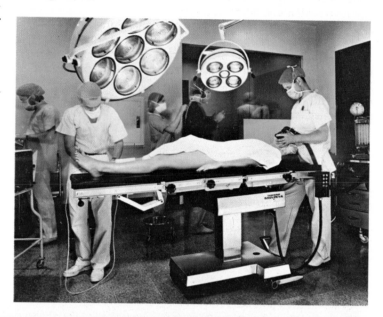

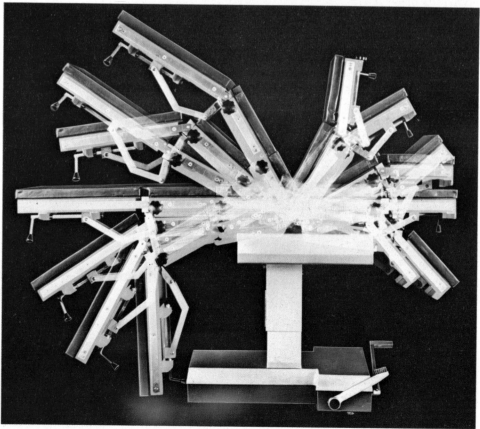

Modern operating tables are capable of a wide range of movements, enabling quick, safe positioning of patients. The five sections of this table can be moved effortlessly and accurately by a hydraulic control system operated from a push-button box. This allows faster responses and greater range of movement than the old method using worm gears and winding handles. The push buttons do not move the table until the operator pumps the foot pedal at the head end, thus preventing accidental movements.

PARTS FOR HEARTS

The heart's job is to pump the entire blood content of the body round and round for a whole lifetime without ceasing. And with heart failure of one sort or another a major cause of death in modern societies, it is hardly surprising that doctors have long sought ways of replacing failing parts of the heart with artificial substitutes.

One of the commonest heart diseases affects the heart valves which control the flow of blood into and out of the various chambers of the heart. When they begin to malfunction, leaks develop between the heart chambers or narrowing of the valves occurs, causing the heart to act less efficiently. The answer is to replace the failing valve with a man-made substitute.

Artificial valves

One of the most commonly fitted *artificial heart valves* is the Starr-Edwards ball valve, in use since 1960. This was named after the US heart specialist Albert Starr and the aircraft engineer M. L. Edwards who designed it. The valve consists of a hollow ball made of a very hard, tough chromium-cobalt alloy (in early models it was made of silicone rubber) held inside a metal cage. The cage and orifice (the opening through which the blood flows) are covered with a cuff of knitted polypropylene to encourage the ingrowth of tissue and discourage blood clots from forming. The cuff also allows the valve to be sewn into position more easily.

Various other designs of artificial heart valve have been developed over the years. With many of them, however, problems have arisen because of defects in design or materials. One of the main troubles was that blood clots tended to form on the valves. Another was that the tissue lining the heart often grew into the orifice, reducing the area through which the blood could flow and preventing the valve from opening and closing properly—a situation that could prove catastrophic for the patient.

The other main problem was that the flow of blood caused a great deal of turbulence which was not present in a natural valve, and this led to destruction of red blood cells and consequent anaemia in the patient.

Recently a valve has been developed that seems to avoid these problems. First tested in July 1976, and known as the St. Jude Medical bi-leaflet valve, it consists of two 'leaflets' that rotate within a ring, upon which a seam-free sewing cuff is mounted. When the valve is closed, the inner edges of the two leaflets meet at the centre of the orifice—they do not lie flat but at an angle of 30 to 35 degrees. As the blood forces the leaflets open, they swing out to a maximum angle of 85 degrees from the horizontal, allowing the blood to pass through three openings, a narrower one at the centre and two larger ones at either side.

The design and angling of the leaflets means that minimal blood pressure is needed to open them, so that they open and close more quickly than other designs. The flow of blood through the valve is smooth, so there is little danger of clots forming or blood cells being destroyed. Because the blood flows through on both sides of each leaflet, the leaflet surfaces are completely washed on both sides during each pump-

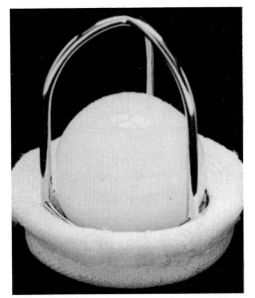

Left: This heart valve consists of a hollow ball made of a very tough alloy housed within an alloy cage. The cuff of knitted polypropylene allows the surgeon to stitch the valve securely in place.

Below: The St. Jude heart valve, stitched into place. This view shows the valve with its two leaflets fully open.

Below right: A view of the valve with its leaflets closed.

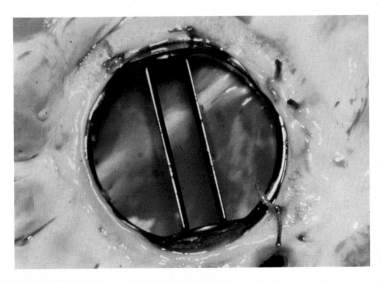

ing cycle; this also helps to prevent clots from forming.

The pivoting mechanism is enclosed by guards; these prevent the ingrowth of tissue which would interfere with the opening mechanism. And the valve is made of machined graphite coated with pyrolytic carbon, a very strong, lightweight and durable material.

Setting the pace

Another lifesaving device is the *artificial pacemaker*, cunningly designed to deliver regular electric stimuli to a heart unable to beat at the correct rate on its own. To explain how it works it is necessary to outline the way in which the four-chambered heart normally operates. Blood from all over the body, deprived of its life-giving oxygen, and laden with waste carbon dioxide, is transported via the veins to end in the right-hand collecting chamber of the heart—the right atrium.

From the right atrium the blood passes through a valve into the right pumping chamber—the right ventricle. The right ventricle contracts regularly to force the blood along the pulmonary arteries to the lungs. In the lungs, carbon dioxide is exchanged for oxygen, and the blood then drains back through the pulmonary veins to the left atrium of the heart. It then passes into the left ventricle, the heart's main pumping chamber. This contracts at a variable rate to drive oxygen-rich blood through the arterial system to all parts of the body. The pumping work itself is done by the heart muscle, or myocardium, which makes up most of the heart's bulk. Unlike other muscles, the myocardium has its own built-in mechanism to maintain the rhythmic heartbeat independently of its nerve connections.

This pacemaker is located in a knot of nerve fibres, the sino-atrial node (S-A node) in the rear wall of the right atrium. It generates a brief low-intensity electrical impulse approximately 72 times per minute in a resting adult. This causes the right and left atria to contract, thrusting the blood into the empty ventricles. The impulse now passes quickly to another specialized knot of tissue, the atrioventrical node (A-V node), near the junction of the atria and ventricles. Here the impulse is delayed for about seven hundredths of a second—precisely the right time to allow the atria to complete their contractions.

Next, conducting fibres fan out through the ventricles to carry the impulse to every muscle fibre so that the ventricles contract within about six hundredths of a second to pump the blood into the arteries. It is damage to these conducting fibres which is the most serious, leading to a condition known as *heart block*—where the heartbeat is slow and cannot be stepped up to meet the demands of exercise. In the late 1950s *artificial pacemakers* were developed—they cannot overcome the underlying disease, but they do force the heart to beat properly and so enable patients to live a normal life.

External pacemakers

External pacemakers were the first type to be developed. Although they have the advantage that batteries can be changed or recharged without recourse to surgery, they suffered from the great drawback that the wires connecting the pacemaker to the heart eventually gave rise to infection where they passed through the skin, and so they were abandoned for permanent use. However, external pacemakers are invaluable in hospitals as a stop-gap measure to tide the patient over a temporary block, which may occur during a heart attack or, much more commonly, during the implantation of a permanent pacemaker.

Implanted pacemakers

Modern pacemakers are implanted entirely within the body, complete with batteries. Although their disadvantage is that they must be implanted surgically and periodically be replaced or recharged, they have replaced the external models for the reasons described.

Each pacemaker consists of two main components: the pacemaker itself, often known as the pulse generator, which

ST JUDE VALVE

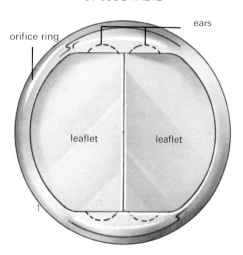

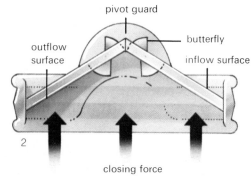

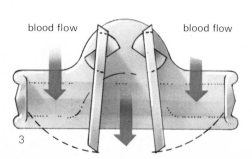

Right: How the St. Jude valve works. Seen from above (1) and the side (2), the two leaflets pivot within the opening, or orifice, fitting snugly at the centre when closed. The pivot arrangement consists of 'ears' at the ends of each leaflet which fit into grooves called butterflies. The pivot guard is to protect the mechanism from becoming clogged by the ingrowth of heart tissue. Blood flow (3) through the open valve is via three separate areas; this keeps turbulence to a minimum and so reduces damage to blood cells.

HEART PACEMAKER PULSE GENERATOR

Right: This is the pulse generator unit of a modern heart pacemaker. It is powered by a lithium iodine battery which, together with an energy-efficient circuit, gives it a life of six years or more; after this time the patient requires a further implant. The tiny quartz crystal enables the pulse rate to be changed quickly and accurately. The parylene coating seals the battery and circuitry from damage by the surrounding body fluids.

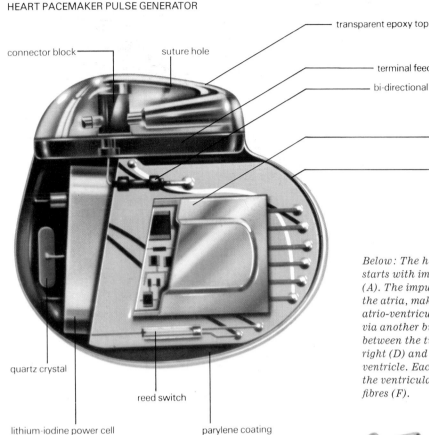

connector block — suture hole

transparent epoxy top

terminal feedthrough

bi-directional zener diode

hybrid circuit

stainless steel case

quartz crystal

reed switch

lithium-iodine power cell

parylene coating

Below: The heart's natural pacemaker system starts with impulses in the sino-atrial node (A). The impulses radiate across the walls of the atria, making them contract. Next, the atrio-ventricular node (B) relays the impulse via another bundle of fibres (C) into the wall between the two ventricles. Here it passes into right (D) and left (E) branches, one for each ventricle. Each branch conveys the impulse to the ventricular muscle via a network of fibres (F).

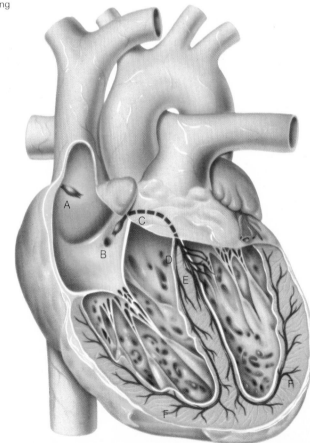

contains the power source and circuitry to produce the right electrical impulse; and the lead, an insulated wire which carries the impulse from the pulse generator to the heart. Some types of pacemaker need two leads. Both pulse generator and leads are generally implanted under local anaesthetic only.

Early pacemakers used mercury batteries, with an average life of only two to three years; these emitted hydrogen gas, so the power unit could not be completely sealed, which meant that body fluid could enter slowly and reduce the life of the battery. A few years ago, it seemed as if nuclear-powered batteries would provide the answer to the problem, but their safety has not been conclusively established, and they have generally been abandoned in favour of the lithium-iodide fuel cell. First developed in Canada in 1973, this is a solid state battery which does not produce any gas, so the power unit can be hermetically sealed against intrusion by body fluids. The battery maintains a steady output which does not decrease markedly until it needs replacing, and will last in some cases for eight years or more.

Demand type

At present, the most commonly fitted pacemaker is the stand-by or demand type, which is particularly useful for the patient who suffers from intermittent heart block. This ingenious little machine incorporates a device that monitors the patient's own ventricular impulse and cuts in only when this natural message fails.

This X-ray clearly shows the pulse generator, and the lead which carries the impulse from the generator to the heart.

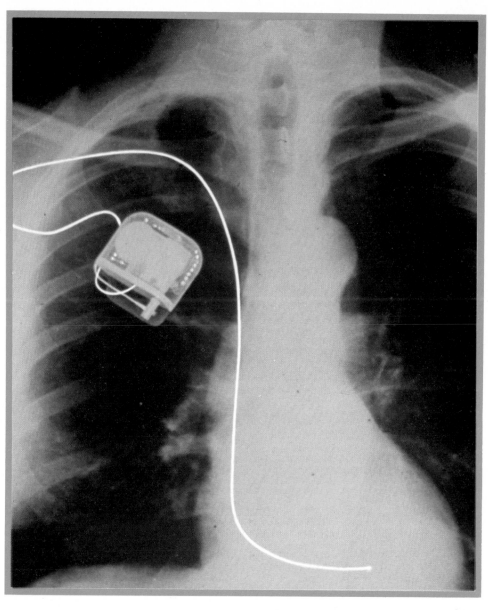

This X-ray clearly shows the pulse generator, and the lead which carries the impulse from the generator to the heart.

Most modern demand pacemakers are programmable—that is, the heart specialist can alter both the pulse rate and the duration of each pulse (the pulse width measured in microseconds). He does this by means of a programmer containing a digital silicon chip circuit controlled by an extremely accurate quartz crystal timing device. This can change the pulse rate or width via an electromagnet which is held about 5cm (2in) above the patient's skin over the buried pacemaker. It sends out digitally coded pulses which are picked up by the pacemaker. The programmer is small enough to be hand-held, yet it can change the pulse rate over a large range—between 30 pulses per minute and 119 pulses per minute in one model—and also the pulse width—between 0.1 and 1.9 microseconds. Moreover, change of rate or width takes only 2.5 seconds.

Phone check
A further, even more remarkable development has been that of a patient follow-up system, by means of which the doctor can check that the pacemaker is working correctly and diagnose potential problems without the patient having to move from his home—a boon for pacemaker wearers who are elderly, infirm or live far away from the surgery. The patient hangs a small transmitting device on his or her chest, dials the number of the surgery or hospital department, gives his name and pacemaker number, then turns the transmitter on and places the telephone mouthpiece over the transmitter's speaker.

When the doctor or his assistant receives the call, he records the patient's name and number, switches on a receiver and places his telephone earpiece on a device called a phone coupler. The unit converts the patient's telephone signal to a digital readout of pacemaker rate and pulse width, and also gives a permanent record in the form of an electrocardiogram—a curve traced on graph paper showing the action of the heart. The couplet can also be connected to other equipment, such as an oscilloscope, for more detailed analysis of the patient's signals.

TEST TUBE BABY

In 1978, after many years of research, the world's first 'test-tube' baby was born, to a woman who had been infertile for many years. The technique of removing an egg cell from the ovary, fertilizing it outside the body, and then implanting it in the womb sounds straightforward, but many obstacles had to be overcome before success was finally achieved.

A common cause of infertility is blockage of a woman's Fallopian tubes, the two ducts that normally provide a passage between the ovaries and the uterus along which an egg can travel. The 'test-tube' baby technique is, quite simply, a means of getting round this blockage.

The launching of a potential human life at the microscopic stage takes place approximately every 28 days in a woman of childbearing age. It begin when the pituitary, a gland at the base of the brain with an important influence on growth and body functions, sends a surge of hormone into the bloodstream. This hormone prompts the rapid ripening and release of one or more of the eggs contained in the woman's ovaries, her two 'sacks' of egg cells which are on each side of the uterus. The ripe egg travels from one or other ovary down the Fallopian tube connecting it with the uterus, which is where, if fertilized, it will grow into a baby.

In the Fallopian tubes

The journey from ovary to uterus takes about three and and half days, and much of this journey is spent passing through the 100 tp 130mm (4 to 5in) long Fallopian tube, the inside of which is only as wide as a piece of thread. The tube is lined with *cilia*, tiny hairs that waft the egg very slowly along its course, ensuring that it travels neither too fast nor too slowly. This is important, because it is while the egg is still in the tube, generally about half-way down, that fertilization takes place.

Within an hour of fertilization occurring, the nuclei of the two cells, the egg and the sperm, fuse together. Another hour later the egg divides into two cells. Then, repeatedly doubling the number of its cells, the egg enters the womb and implants itself in the lining, where if all goes well it will stay for the rest of the nine months of pregnancy.

As the Fallopian tubes play such a vital part in this process, the slightest damage to them can break down the whole reproductive mechanism, by making it impossible for egg and sperm to meet. Roughly one in ten couples in the western world is unable to bear children, and in about one third of these the cause is blocked or damaged Fallopian tubes. The damage is usually caused by infection or occasionally by abdominal surgery. Sometimes an operation to try to unblock the damaged tubes is successful, but most such women remain childless.

In the early 1960s an idea developed in the mind of Patrick Steptoe, a gynaecologist and obstetrician at a hospital in Oldham, Britain, who had seen many such cases among the patients attending his clinics. It occurred to him that the blockage might be bypassed in a fairly straightforward way if four steps could be achieved. These were, first, to remove an egg from the woman surgically and, second, to fertilize it in the laboratory with sperm from her partner. (This is not done in a test tube at all, in fact, but in a shallow vessel called a *petri dish*. The correct technical term is *in vitro fertilization*, but 'test-tube' fertilization is the term that has stuck with the research since its earliest days.) The third step was to keep the egg alive in the laboratory for a few days until the beginnings of an embryo were achieved, and the final step was to implant this embryo directly in the woman's uterus, to grow there as in any other pregnancy. In essence, this is the 'test-tube' baby technique as it was ultimately developed after years of research.

Removing the egg

Patrick Steptoe already had at his disposal an instrument for examining the ovaries and Fallopian tubes, called a *laparoscope*. This is a kind of endoscope (see page 80) which can be inserted through a small cut in the abdomen, allowing the surgeon to look inside the abdominal cavity. He had pioneered the use of this instrument to find out why some women were infertile. It was clear to him that it could also be used, together with a needle-like suction rod, to remove eggs from an ovary of a woman with blocked Fallopian tubes.

In the mid-1960s Steptoe found in Dr Robert Edwards, a Cambridge physiologist who had spent years researching the problems of laboratory fertilization, the very person he needed to help him take his idea further. Dr Edwards had learned that to repeat in the laboratory the apparently straightforward way fertilization takes place naturally required very much more than mixing egg and sperm together.

Age for maturity

For fertilization to occur, the egg—about the same size as a full stop on this page—has to be at exactly the right stage of

This aspirator (right) is a tube 1.3mm in diameter, through which the egg is sucked out from the potential mother.

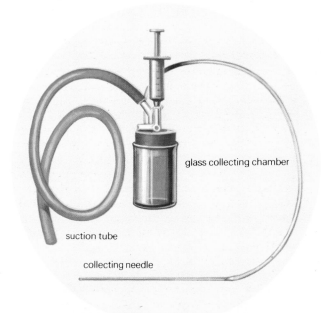

glass collecting chamber

suction tube

collecting needle

maturity when it meets the sperm. The timing of its removal from the ovary therefore has to be perfect. The maturing process continues during the journey down the Fallopian tube, where special secretions are present. Dr Edwards discovered how to reproduce these conditions in the laboratory. He created a mixture of nutrient chemicals, kept at body temperature, that allowed the ripening process to continue.

By 1966, Edwards knew how to recognize under the microscope the exact moment when the egg is ripe and capable of being fertilized: it changes its appearance slightly when it is ready to fuse with the male cell. However, he was still working on a similar problem with the spermatozoa, 20,000 times smaller than the eggs. They too undergo an obscure change, known as *capacitation*, during their journey up the oviduct. Unless 'primed' in this way they will not fuse with the egg.

First breakthrough

The first real breakthrough came in February 1969, when Steptoe, Edwards and another Cambridge scientist announced in the scientific journal *Nature* that for the first time they had achieved the fertilization of several human

eggs outside the body. But further difficulties were experienced before they were able to achieve a successful pregnancy. For example, it was found that artificially ripening eggs in the mother ('super ovulation') did not allow proper development of the fertilized egg, and it added to the already great difficulties of persuading the uterus to accept the potential embryo. So, although, it is more difficult, a naturally ripened egg is taken from the mother at just the right moment in the natural cycle. Timing is obviously crucial, and the techniques developed demand that a team of theatre nurses be on constant standby, while hormone levels in urine and blood samples are monitored round the clock to watch for the hormonal surge that shows that the egg is ready.

Fertilization

When the surge occurs, sperm from the male partner is collected and prepared, and 18 to 22 hours after the onset of the surge an egg is collected from the woman by laparoscopy. Egg and sperm are then mixed in a dish containing a culture medium, and kept at body temperature. When fertilization

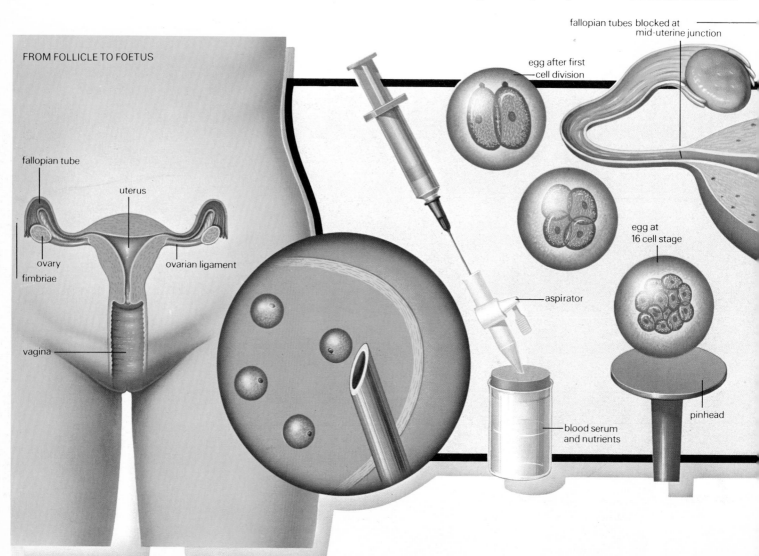

FROM FOLLICLE TO FOETUS

fallopian tube
uterus
ovary
ovarian ligament
fimbriae
vagina

fallopian tubes blocked at mid-uterine junction
egg after first cell division
egg at 16 cell stage
aspirator
pinhead
blood serum and nutrients

takes place, the egg starts to divide and grow, and after two or three days is placed in the uterus. Further hormone monitoring is maintained to find out whether pregnancy has been established.

In vitro fertilization could be used to help couples with problems other than blocked Fallopian tubes. As relatively few sperm are needed to fertilize an egg in a laboratory dish, a man with a low sperm count who would be otherwise infertile could father a child in this way.

The technique could be further extended to help women who cannot conceive because of damage to their ovaries, preventing the release of eggs. They could instead receive an egg from a donor, to be fertilized by their partner. The women would then be able to carry a pregnancy and give birth in the normal way.

Test-tube babies are now a reality. But before the *in vitro* fertilization method leaves the experimental stage and is put into widespread use, the doctors and scientists involved still have to improve their understanding of why so many attempts fail, and adjust their methods to ensure success in future.

Any woman hoping to conceive by the Steptoe technique is monitored round the clock to catch the hormone surge, which is the best time to take the egg from the ovary. To remove the egg, the surgeon makes an incision in the abdominal wall to insert a laparoscope and see the ovary. Through another incision he inserts the hollow needle of an aspirator to gently suck out the minute egg. The egg is then transferred to a dish of nutrients kept at body temperature and the father's sperm is introduced. Soon one of the sperm penetrates the egg. After about eight hours an egg chromosome and the sperm fuse into a single cell that divides and subdivides over the next two to three days until there are eight or 16 cells within the egg. At this point, the fertilized egg is implanted in the womb very simply via the cervix through a narrow polythene tube or cannula. Pregnancy can then develop normally as the egg becomes attached to the thickened uterus lining, works its way in and grows into a healthy foetus.

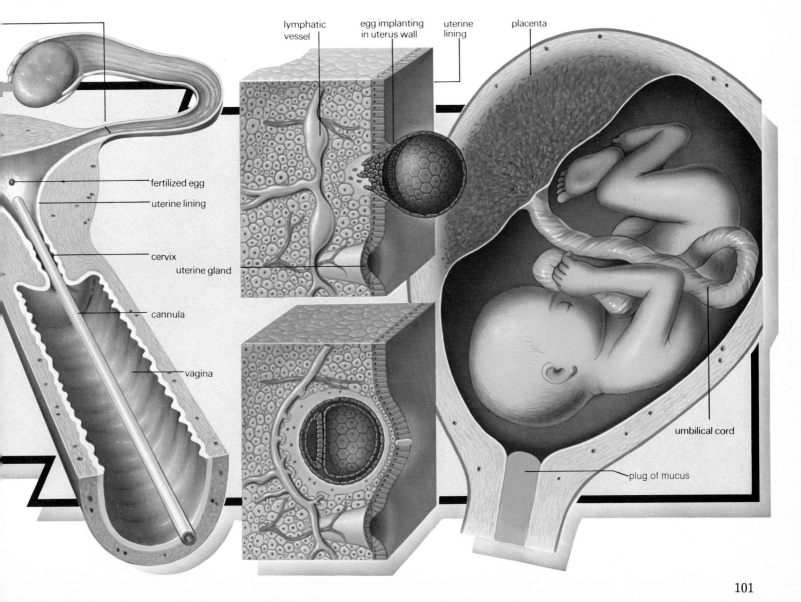

fertilized egg

uterine lining

cervix

uterine gland

cannula

vagina

lymphatic vessel

egg implanting in uterus wall

uterine lining

placenta

umbilical cord

plug of mucus

MILITARY

ARMOURED VEHICLES

The problem of attacking enemy trenches, protected by barbed wire and defended with machine guns, was solved by a new type of vehicle—one carrying weapons, immune to bullets and capable of crossing any terrain. An innocuous name was needed for security reasons. Since the first ones vaguely resembled water tanks, they were called 'tanks' and the name stuck.

The first tanks were developed almost simultaneously in Britain and France. A prototype was built in September 1915 and one year later the first British tanks went into action, on the Somme, in France. By the end of the war tanks had scored several major successes and Britain had already produced 2600, while France had produced 3870, mostly of a lighter type.

Between the end of World War I and the outbreak of World War II tank design made considerable progress, transforming them from ponderous assault machines capable of no more than 8 to 10km/h (5 to 6mph) into versatile fighting vehicles capable of 48km/h (30mph).

The German Leopard 2 typifies the modern approach to tank design with its immensely powerful engine, high-energy shock absorbers and heavy armour. It is also highly mobile and has heavy fire power.

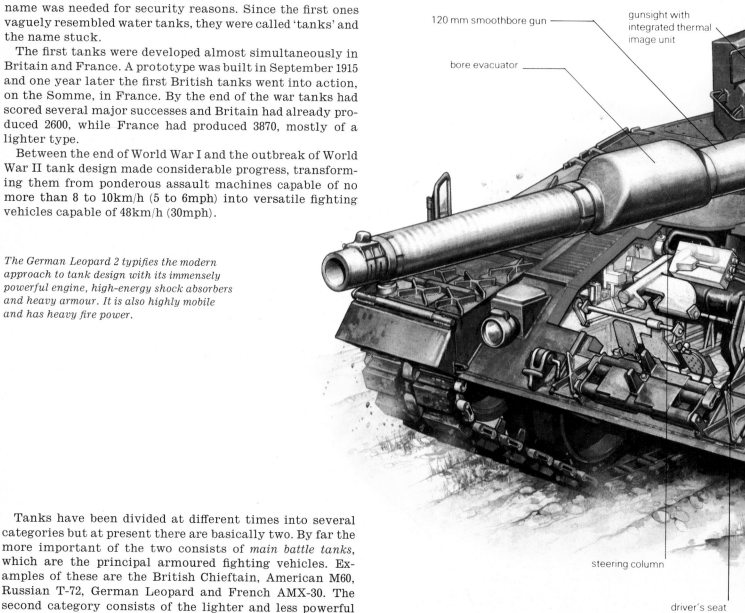

LEOPARD 2

120 mm smoothbore gun

gunsight with integrated thermal image unit

bore evacuator

steering column

driver's seat

Tanks have been divided at different times into several categories but at present there are basically two. By far the more important of the two consists of *main battle tanks*, which are the principal armoured fighting vehicles. Examples of these are the British Chieftain, American M60, Russian T-72, German Leopard and French AMX-30. The second category consists of the lighter and less powerful *reconnaissance tanks* which are exemplified by the British

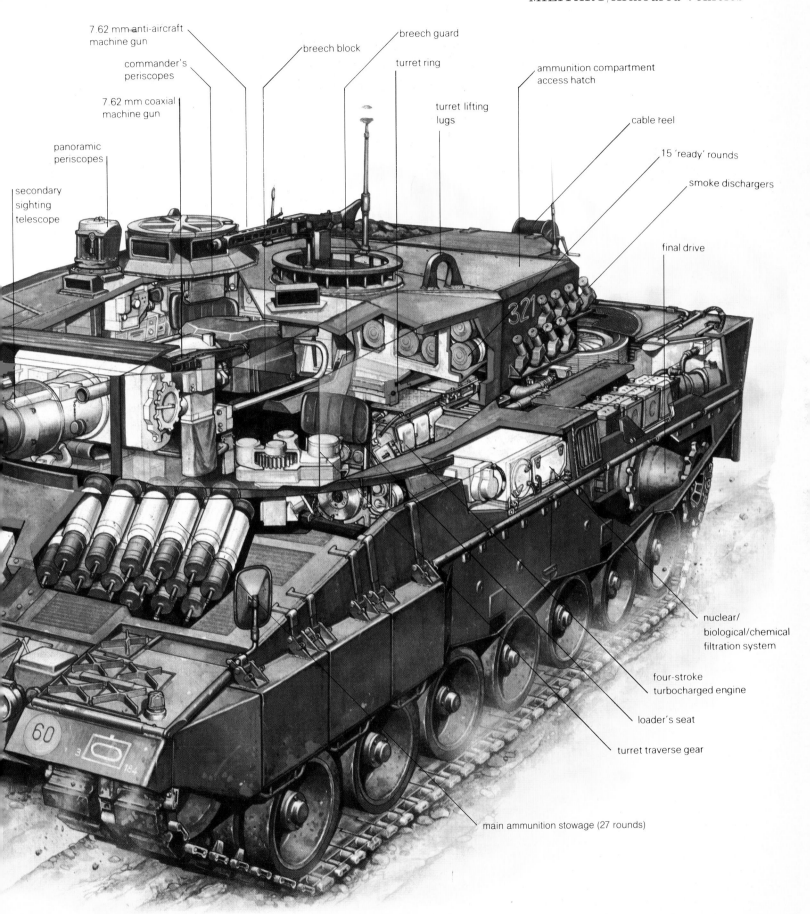

7.62 mm anti-aircraft machine gun

commander's periscopes

7.62 mm coaxial machine gun

panoramic periscopes

secondary sighting telescope

breech block

breech guard

turret ring

turret lifting lugs

ammunition compartment access hatch

cable reel

15 'ready' rounds

smoke dischargers

final drive

nuclear/biological/chemical filtration system

four-stroke turbocharged engine

loader's seat

turret traverse gear

main ammunition stowage (27 rounds)

The Swedish Stridsvagn 103 is designed to be small in height. There is no gun turret, so the 105mm gun is fixed rigidly to the hull: it is elevated by tilting the hull on the advanced suspension. This low profile approach means that the tank presents a particularly small target. Best use of this is made when the tank operates in the 'hull down' position—when it is placed on a reverse slope so that the hull is under cover, but the gun is still clear and usable. This technique is not unique to the Stridsvagn—many heavily armoured tanks operate in this position—but the Stridsvagn also has the advantage of a retractable bulldozer blade fitted to its front, which enables it to dig itself in.

Scorpion, American M551 Sheridan and Russian PT-76. There are also anti-aircraft tanks for defending tank units against attack from the air, and other vehicles based on tank chassis, which are also sometimes called tanks. These include recovery vehicles for winching or towing disabled tanks, and bridge-laying vehicles.

Since tanks are, in principle, mobile weapon platforms, their design centres on their armament: above all else, this must be capable of destroying enemy tanks. This requirement has been met by guns ranging in calibre from 75 to 125mm. The *armour piercing discarding sabot*, or APDS, type of projectile, can penetrate armour whose thickness is approximately four times the calibre of the gun. Many tank guns also fire an entirely different type of *high explosive anti-tank*, or HEAT, projectile which is capable of penetrating even thicker armour. In addition, tank guns also fire conventional, high-explosive shells and they carry secondary armament in the form of a rifle-calibre machine gun mounted beside the main armament. Some can also fire guided missiles.

To enable tanks to fire on the move, guns are stabilized by two gyroscopes which keep the gun at a given elevation and bearing however much the tank is thrown about.

In most cases, tank guns are mounted in turrets which provide all-round, 360-degree traverse. The turrets are generally manned by three men: the gunner, to aim and fire; the second crewman or loader, to feed the guns; and a commander, to decide on the target. The crew of a tank is completed by a driver who is located in the hull, ahead of the turret. The rear portion of the hull is taken up by the engine and transmission.

A tank is most likely to be hit at the front, so the armour here is very thick—120mm (4.7in) or more. Elsewhere, especially at the rear, the armour is thinner to help keep the total weight down. This is important, because the mobility of a tank depends greatly on its weight—or rather the ratio of the power developed by the engine to its weight. For a given engine power, the power-to-weight ratio depends largely on the thickness of the armour, since this makes up about half of the total weight of a tank; so to make a tank fast means sacrificing some of its protection. The British Chieftain is an example of a tank with a fairly poor power-to-weight ratio, easily bettered by the French AMX-30: the

British view is that sturdy armour is the best way to ensure a tank's survival, even if this does make it less mobile.

Sloping the front of a tank, which deflects high-velocity ammunition, provides additional protection, but sloping armour is still vulnerable to HEAT type chemical energy warheads. The answer is a composite armour, with two layers of steel sandwiching a layer of plastics material.

Tanks ride on what are effectively elongated wheels, called *tracks*. These spread the weight of the tank over a large area, reducing the ground pressure, and enabling the tank to operate across country. Each track is made up of a hundred or so pin-jointed steel links.

Tracks are made as wide as possible. However, they cannot be increased strictly in proportion to the weight of the tank so the ground pressures range from about $0.8kg/cm^2$ (12psi) down to only $0.35kg/cm^2$ (5psi) for light tanks such as the Scorpion, which can move over soft ground difficult even for men to cross on foot.

The weight of a tank is transferred on to the tracks, by four to seven road wheels on each side. The wheels are solid rubber tyred and in most cases are independently mounted on single arms pivoted in the hull, and independently sprung. The Swedish S-Tank has hydro-pneumatic suspension. This is adjustable, permitting the raising and lowering of the tank within certain limits, and even tilting it sideways or fore-and-aft. The S-Tank uses this to aim its gun.

The characteristics of the suspension have a major influence on the cross country speed of a tank, which is limited by the amount of pitching and bouncing that the crew can withstand rather than the power of the engine. Even the best suspensions cannot prevent the speed of tanks being considerably lower over rough ground than on roads. To minimize the pitching and bouncing when moving over rough ground, a tank should be as long as possible, but the longer it is, the more difficult it is to steer. A compromise between these requirements leads to a length to width ratio of about 1½:1.

Manoeuvrability is not just a question of speed: a tank must also be handy at turning. The modern technique is called regenerative steering. In this system, driving power is subtracted from one track and transferred to the one on the opposite side through a differential geer; the system enables a tank to slew around on a point.

BOMB DISPOSAL UNIT

The problems of bomb disposal did not end with World War II. If anything, they have increased. Thousands of anti-aircraft shells are still being discovered, as builders dig down to uncover them. Mines, demolition charges, stores of bombs and explosives hidden during the war and then forgotten emerge regularly. Decades after the war, bomb disposal squads in major European cities are dealing with over a thousand explosive items every year.

The World War II German bomb fuse was electrical—as it left the aircraft a charge of electricity was sent into the fuse, and stored in a capacitor. When the bomb landed, the impact would close a switch, allowing the current to flow into a detonator and fire the bomb. Occasionally, the bombs failed to work because the switch failed to operate properly. But disturbing the bomb—turning it to get at the fuse, or unscrewing the fuse itself, could cause the switch to close. So the first piece of technical equipment to enter the bomb disposal armoury was an electric lead which clamped on to the fuse and safely discharged the fuse capacitor—making the fuse completely inert and safe to remove.

As well as the electrical switch, bombs also had a clockwork delay timer—this could often be jolted back into action by the activity of the disposal squad. Once a bomb had started ticking, the best thing to do was to evacuate the site, either until the ticking had stopped, or until the bomb exploded.

To check for ticking, one member of the disposal squad would listen with a stethoscope as soon as the bomb had been uncovered.

There are many ways, however, of jamming a clock mechanism so that ticking could not start—by using an extremely powerful magnet, by 'freezing' the mechanism with carbon dioxide, or by filling it with a quick-setting plastic compound.

As the bomb disposal men became more proficient, bomb designers started to incorporate booby traps to prevent the fuse being removed without activating it. Each new development meant that a new method of disabling the bomb had to be carefully worked out.

Urban warfare

Today, the bomb disposal business has entered a new phase as the terrorist and urban guerilla have appeared on the scene. Indeed, so many and varied are the devices that have appeared that a new term has been coined—*explosive ordnance disposal* (EOD).

At first, the amateur terrorist bombers made many mistakes. They assembled bombs incorrectly so the devices

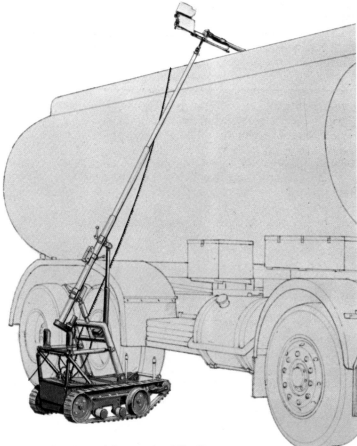

Two applications of the amazing Wheelbarrow
—a one-machine bomb disposal squad.
Left: The handling grab is moving a suitcase,
thought to contain explosives, to safety.
Above: The telescopic boom enables a
television camera to examine the top of a
petrol tanker.

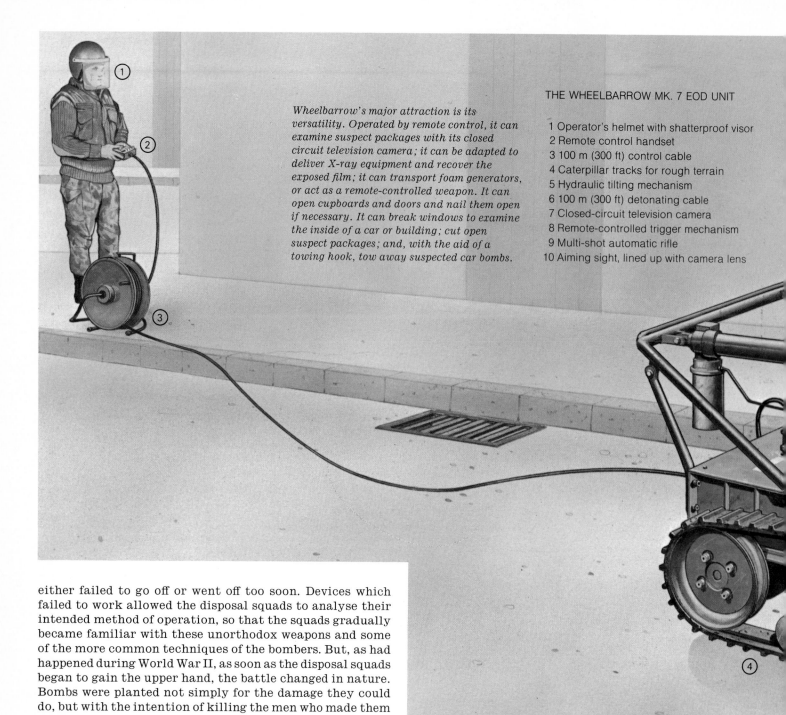

Wheelbarrow's major attraction is its versatility. Operated by remote control, it can examine suspect packages with its closed circuit television camera; it can be adapted to deliver X-ray equipment and recover the exposed film; it can transport foam generators, or act as a remote-controlled weapon. It can open cupboards and doors and nail them open if necessary. It can break windows to examine the inside of a car or building; cut open suspect packages; and, with the aid of a towing hook, tow away suspected car bombs.

THE WHEELBARROW MK. 7 EOD UNIT

1 Operator's helmet with shatterproof visor
2 Remote control handset
3 100 m (300 ft) control cable
4 Caterpillar tracks for rough terrain
5 Hydraulic tilting mechanism
6 100 m (300 ft) detonating cable
7 Closed-circuit television camera
8 Remote-controlled trigger mechanism
9 Multi-shot automatic rifle
10 Aiming sight, lined up with camera lens

either failed to go off or went off too soon. Devices which failed to work allowed the disposal squads to analyse their intended method of operation, so that the squads gradually became familiar with these unorthodox weapons and some of the more common techniques of the bombers. But, as had happened during World War II, as soon as the disposal squads began to gain the upper hand, the battle changed in nature. Bombs were planted not simply for the damage they could do, but with the intention of killing the men who made them safe.

So nowadays a bomb disposal expert is likely to be protected by an 'EOD Suit'. Made of ballistic nylon this is designed to offer as much protection as possible should the bomb detonate, and still allow the operator some measure of mobility.

As a rule, the first step in examining a suspected bomb is to X-ray it. This reveals the internal mechanism, the arrangement of component parts, and the layout of electrical wiring. The first X-ray equipment used for this task was extremely cumbersome, requiring a truck to move it into position. But the miniaturization made possible by modern electronics has led to portable equipment. The operator can

also use an explosives detector called a 'sniffer' to check that the device really is a bomb, and not a time-wasting hoax.

If the bomb relies upon an electric battery to provide power for the detonator (most bombs incorporate a battery), then freezing the bomb will render the battery inert. The bomb can then be safely dismantled. Alternatively, a disruptor aimed at some sensitive part of the bomb discovered by X-ray examination can be fired at short range. Or the container can be pumped full of plastic foam, short-circuiting the electric circuits and rendering any mechanical devices inert. All these, and several more techniques which

are not made public, are available to the EOD disposal team.

But as terrorists in the UK started to incorporate anti-lift, anti-open, anti-disturbance switches into the bombs it became obvious that a remote approach to disposal had to be devised and developed—'Wheelbarrow' was one answer.

'Wheelbarrow' is the name of a small, remotely-controlled tracked vehicle with an articulated arm carrying a closed-circuit television camera and floodlight. This allows bombs to be examined at a safe distance. Wheelbarrow can lift a bomb, so that the camera can examine it more closely, or to move it. It can also fire a special 'disruptor' projectile at a

bomb. This breaks up circuits and fuses so quickly that the bomb has no time to operate.

At present, the British Army is testing a much improved model of Wheelbarrow known as 'Marauder', Marauder has longer articulated tracks to give better travel over rough surfaces. The machine can even climb up and down stairs.

The battle will continue. So long as someone is prepared to construct and plant a bomb, there will be an explosive ordnance disposal operator to take on the task of rendering it safe. And he will have an ever-increasing array of techniques available to help him.

BULLETS

Effective small arms ammunition is crucial to any army. Yet despite the advances in other military hardware, the four main components of a conventional metallic cartridge (the case, primer, propellant charge, and bullet) saw few developments from about 1886 until World War II.

Then in Germany, a new development in rifle ammunition occurred that was to change post-war design and development considerably. Germany first recognized that infantry fighting ranges were no longer about 900m (1000yd) or more, for which the first metallic cartridges were designed, but were, in most instances, about 350 to 550m (390 to 600yd) only. For such reduced range, and for use in a new type of automatic rifle, known in Germany as the 'Sturmegewehr' (assault rifle), Germany produced a new 7.92mm cartridge with a shortened case. This new short-cased cartridge was ideal for the new class of rifle; it was lighter and produced less recoil. After World War II several countries, including the USSR, Britain and the USA, introduced similar cartridges.

Rifle ammunition

Several development trends in rifle cartridges are now apparent. Research conducted largely in the USA focused on the problems of 'hit probability' with the rifle fired by an ordinary soldier—how likely a soldier is to hit a human target. This work included an analysis of battle casualties, and the result confirmed that hit probability fell off sharply once the engagement range increased beyond about 100m. This investigation in the early 1950s resulted in the first of a series of experimental small-calibre, high-velocity cartridges. In the US experiments, small calibre—5.56mm (0.223in)—was combined with a relatively long case—about 45mm (1.77in)—to produce an extremely lethal, high-velocity cartridge. The bullet was reduced to nearly a third of the original weight of the 7.62mm NATO cartridge, so that it weighed only 3.5g (55 grains), compared with the 9.3g (144 grains) of the NATO ball bullet.

This reduction in weight helped to increase velocity, but at the expense of range which was less important. The muzzle velocity of the new bullet was about 970m/sec (3100ft/sec), a considerable advance on the 820m/sec (2700ft/sec) of the 7.62mm NATO bullet. But most importantly, the new cartridge could be fired from a new, light-weight automatic rifle. Such a rifle was then developed in the USA, originally called the Armalite, and eventually known by its US Army designation of M16.

Cartridge cases

It should be stressed that, although representing a considerable break with the past in terms of calibre and weight, both the Soviet and the American-NATO cartridges are conventional in design: they comprise case, primer, propellant and bullet. But extensive work being done to reduce the number of cartridge components, to save weight and cost, and conserve strategical raw materials may change all this. The commonest and best favoured approach is the *consumable-cased cartridge*, in which the case is made of solid propellant.

The bullet is secured in a solid cylinder of propellant, fashioned in the form of a normal cartridge case. In the base of the cylinder is secured a pellet of priming composition. When the primer is struck by the firing pin, the solid propellant case ignites and burns, being totally consumed in the process and produces a hot, expanding gas which expels the bullet through the rifle barrel.

Major problems encountered with consumable-cased cartridges relate to breech blocking, overheating and safety. With a consumable-cased cartridge, the heat can be transferred only to the weapon itself, causing serious problems of weapon heating.

Most cartridge cases are made of brass, though in some countries steel cases are standard. Aluminium has been tried, to save weight, but without much success. Similar savings in weight and in strategic raw materials of cartridge cases (mainly copper) were expected from the use of plastic instead of brass or steel but, as yet, this expectation has not been realized for ball ammunition, although plastic-cased training ammunition (blanks) and grenade propelling ammunition are used in several countries.

The original solid lead projectiles, common when metallic cartridges were first introduced, gave way in the 1880s to composite bullets having metal envelopes with, usually,

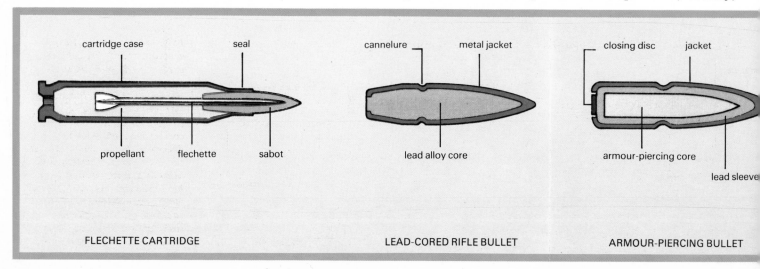

FLECHETTE CARTRIDGE   cartridge case   seal   propellant   flechette   sabot

LEAD-CORED RIFLE BULLET   cannelure   metal jacket   lead alloy core

ARMOUR-PIERCING BULLET   closing disc   jacket   armour-piercing core   lead sleeve

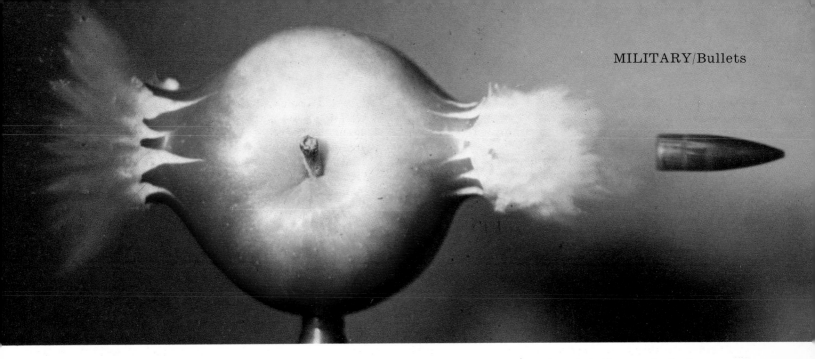

lead alloy cores. Round nosed bullets gave way to pointed or *spitzer* bullets at about the turn of this century, and until recently little basic change occurred in ball bullet design.

An efficient method of causing a bullet to transfer its energy quickly to the target has come from Germany. Experimental bullets with *Löffelspitz* have recently been produced. Such bullets have the area near the tip scooped out, as if by a spoon—hence *Löffel*, the German word for spoon. Another method is the asymmetrical bullet which is designed to tumble more readily than a normally shaped bullet when the target is struck, thus increasing wound effect.

Pistol calibre ammunition

The ammunition for revolvers, self-loading pistols and sub-machine guns has changed less than rifle calibre ammunition. The main development in pistol calibre ammunition is now in the bullet design where there is a new problem. For anti-terrorist use in urban areas, the target may be fleeting and there will not be opportunities for a second shot. Accuracy and killing power are of prime importance. A further important consideration is that, especially when hostages are involved, or with bystanders at risk, the bullet should ideally not be able to continue and injure or kill the innocent.

A bullet passes through an apple virtually unchecked. A high-velocity bullet is particularly lethal because its shock waves damage tissues beyond the immediate wound.

Some designs—many originating in West Germany—have been produced to cope with this requirement. They are known under the generic term of *effect geschoss*. The basic effect geschoss design has a nose cavity in the bullet, covered with a thin metal shroud so that the bullet seems to be full jacketed, or filled with a plastic plug. The intention is that the cavity causes the bullet to expand upon impact. Effect geschoss bullets are lighter than normal ball bullets and have higher velocity.

Normal bullets are ineffective in stopping cars, and several alternative designs exist to cope with such targets. In one form, the bullet has a hard steel core inserted as a separate component. In a more recent design, the bullet has a clad steel envelope with a normal lead core, but the envelope at the nose is specially thickened to form an armour-piercing cap. In the US, it was discovered that all-steel bullets coated with Teflon had special piercing qualities. The Teflon provided lubrication and, therefore, improved penetration significantly.

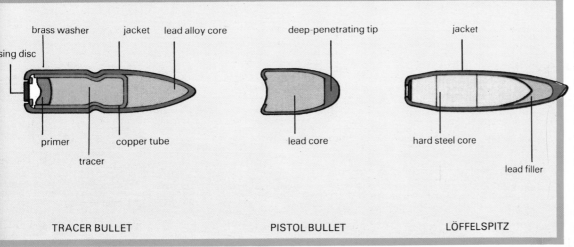

TRACER BULLET — brass washer, jacket, lead alloy core, sing disc, primer, tracer, copper tube

PISTOL BULLET — deep-penetrating tip, lead core

LÖFFELSPITZ — jacket, hard steel core, lead filler

Left: Bullet design varies, according to the type of gun used to fire it, and the effect required. Among the designs shown is the flechette cartridge —this has a thin steel arrow (the flechette) mounted in a carrier or sabot which is discarded during flight. Because the flechette is extremely light in weight, it travels at a very high velocity—about 1400m/sec (4600ft/sec). Although lethal at short range, the flechette rapidly loses accuracy. Also shown is the scooped-out area at the tip of a Löffelspitz bullet.

CRUISE MISSILES

Costing a mere tenth of a traditional intercontinental ballistic missile, the cruise missile has blurred the traditional distinction between nuclear and conventional weapons, and has provided a heated talking point both for arms limitation negotiators and for peace bodies. Yet the idea of a cruise missile is by no means new.

A cruise missile is simply a pilotless aircraft with its own air-breathing jet engine. It has a guidance system to deliver a warhead at the end of its flight, generally at subsonic speeds and, unlike a ballistic missile, entirely in the atmosphere. As a concept the cruise missile dates back to experimental British and American radio-controlled flying bombs of World War I.

The first in service (1976) of the new US cruise missiles was the McDonnell Douglas Harpoon. Turbojet powered and radar homing, it has a range of about 100km (62 miles). It has a guidance system of a radar altimeter for sea-level flight, a mid-course correction unit and a target-detection radar, and can be fired from 533mm (21in) torpedo tubes. Harpoon's radar can pick out two widely separated targets and head for the larger one.

But the anti-ship SLCM (sea-launched cruise missile) needs long-range target data from a third vehicle. The line of sight at sea does not exceed 50km (31 miles) and the radar's range is about the same—hence the supersonic Harpoon's short range. Furthermore, the mid-course unit has a drift of 0.2 radian per hour of flight. At 500km (310 miles) the error could be as much as 40km. The correction has to come from a vehicle, such as an aircraft, able to identify the target and communicate with the missile or its launch platform. If the latter is a submerged submarine, it cannot be contacted or communicate itself with the missile.

Harpoon's Teledyne turbojet, first of a revolutionary breed of engines, has an operational life of less than 15 minutes and, made mainly of cast metal, costs only about $15,000. It weighs about 45kg (less than 100lb) but can generate a thrust of 300kg.

The long-range, land-attack SLCM and the ALCM (air-launched cruise missile) have in common the terrain contour-matching technique or TERCOM system of the McDonnell Douglas Astronautics Company. This system confers pinpoint accuracy on the new generation of cruise missiles. Computer-controlled, the missile navigates by comparing the height of the land beneath it with satellite-produced digital maps of the ground contours stored in its memory—(see page 116).

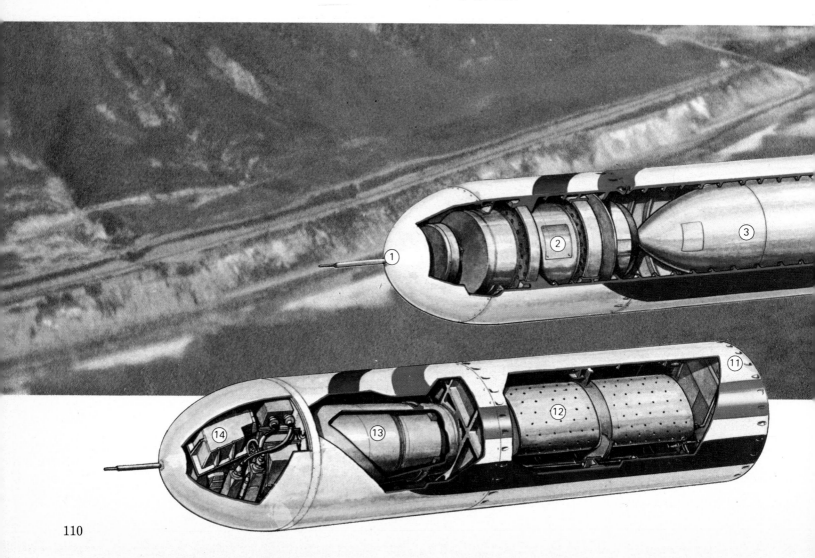

TERCOM works best over hilly ground; not over water. The landfall map, if required, is therefore made wide enough (up to 9.6km/6 miles) for the missile not to miss after its preset guided flight over the sea under an inertial guidance system which is relatively inaccurate. In theory TERCOM makes the SLCM and ALCM accurate to within 9m (30ft). But in practice mapping errors or unsuitable terrain can increase the margin to about 100 yards.

General Dynamics named its sea-launched cruise missile *Tomahawk*, and although able to carry the W-80 thermonuclear warhead, it was made so compact that it could be fired, like the Harpoon, from torpedo tubes. In this form, designated BGM-109, Tomahawk is sealed in a steel capsule. After being fired underwater, the capsule is discarded, the boost motor behind the tail fired and the missile programmed into a steep climb at about 90km/h (56mph). To transform the compact underwater projectile into an efficient aerodynamic vehicle, the boost motor drops off, and wings and tail extend as BGM-109 climbs to cruise level.

Tomahawk quickly became an exceptionally versatile missile. In addition to the strategic land attack version, it was built in an anti-ship version with a modified Harpoon missile, large conventional warhead and active radar guidance to home in on hostile ships. A third and quite different model is AGM-109, the air-launched model, also called TALCM (Tomahawk ALCM). Without either a surrounding capsule or a boost motor, it has been carried by US antisubmarine aircraft which do not normally have much capability against surface targets, at least not against distant cities.

Yet another version, perfected in 1977, is GLCM, the ground-launched cruise missile. Designated BGM-109B, this emerged without much fuss and was an obvious and predictable development. All existing strategic missiles and aircraft in the West, except those in submarines, are launched from vulnerable fixed bases which automatically become targets for Soviet counter-force missiles. For example, there are just two bases in Britain for US F-111 swing-wing bombers: RAF Upper Heyford, Oxfordshire, and RAF Lakenheath, Suffolk. But a mobile cruise missile changes the picture completely. In any time of crisis it would be dispersed to some distant location. No Soviet missile could be targeted on it, because its location would not be known.

In 1980 the British government announced that from December 1983 a force of 160 GLCMs would be based in Britain. Like all GLCMs they will be deployed aboard truck launchers able to drive unobtrusively along public highways to preselected but secret firing locations.

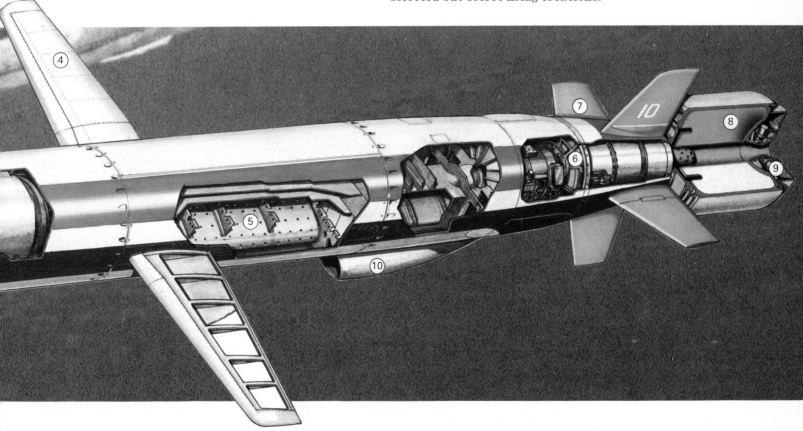

The anti-ship version of the Tomahawk cruise missile and the alternative land attack warhead (11). The missile is small enough to be launched from a submarine torpedo tube. About 10m (30ft) out of the tube, booster rockets (8) fires and jet tabs (9) steer Tomahawk out of the sea at 88km/h (55mph); tail fins (7) spring out to roll it the right way up. After a few seconds the booster is jettisoned and wings (4) extend. The air scoop (10) and turbofan engine (6) take over. The missile has a 450kg (1000lb) high explosive warhead (3) and 550kg of fuel (5); the land attack version has a 120kg (270lb) nuclear warhead (13) and extra fuel (12). Both versions have an airspeed indicator (1) linked to a highly sophisticated navigation system (2 and 14).

INFANTRY WEAPONS

The infantry are the most flexible and adaptable of land forces, capable of going where armoured forces and even artillery cannot penetrate. But the idea of a 'foot' soldier, equipped only with his rifle, is an old one. Today's infantryman can call on a wide range of guns, ranging from pistols to anti-aircraft guided missiles.

The most outstandingly successful rifle since World War II has been the Soviet AK-47 which was designed by Mikhail Kalashnikov and fires 7.62mm ammunition with acceptable accuracy. When a bullet is fired, the pressure of released gases which drive it down the barrel is as great backwards as forwards. In the AK-47, and most other types of automatic rifles, this backward pressure is used to push the bolt back, eject the empty bullet case and allow a new round into position. This can then be pushed into the barrel by the returning spring-loaded bolt and fired. The AK-47 is fed by a 30-round, detachable magazine and can be used to fire single shots as well as bursts of fire.

The AK-47 has not been without rivals. The most significant of these is the Armalite, which was designed some 20 years ago by the American Eugene Stoner. One of the chief difficulties in equipping an infantryman is to keep down the weight he has to carry into action. The Armalite AR-15 solved some of the problems by being designed to fire a lightweight bullet at such high velocity that it was as lethal and accurate as heavier rounds. Lightweight materials such as nylon, plastic and metal alloys were used in manufacturing the AR-15 and it overcame initial teething problems to prove itself reliable. New rifles with smaller, lighter rounds are now entering service.

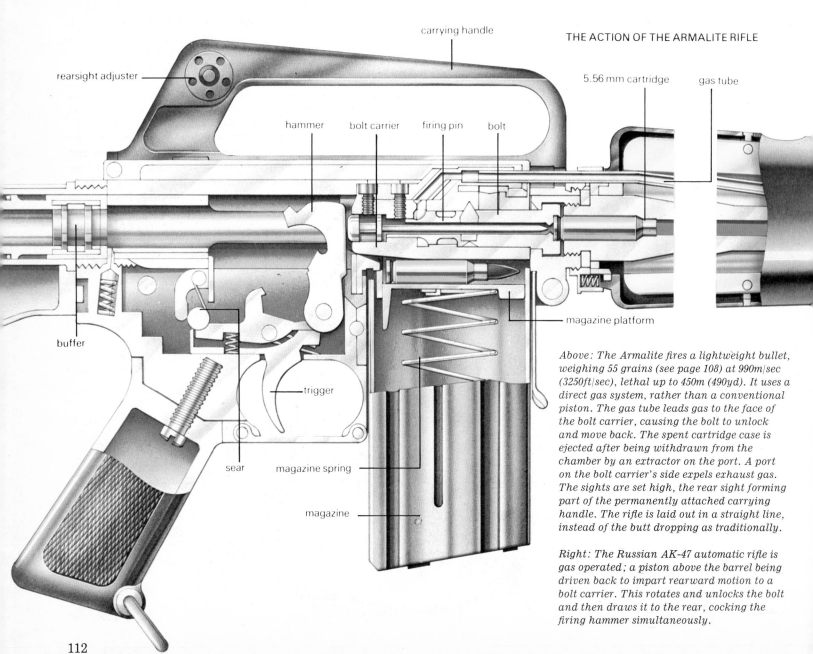

THE ACTION OF THE ARMALITE RIFLE

carrying handle

rearsight adjuster

5.56 mm cartridge

gas tube

hammer — bolt carrier — firing pin — bolt

buffer

magazine platform

trigger

sear — magazine spring

magazine

Above: The Armalite fires a lightweight bullet, weighing 55 grains (see page 108) at 990m/sec (3250ft/sec), lethal up to 450m (490yd). It uses a direct gas system, rather than a conventional piston. The gas tube leads gas to the face of the bolt carrier, causing the bolt to unlock and move back. The spent cartridge case is ejected after being withdrawn from the chamber by an extractor on the port. A port on the bolt carrier's side expels exhaust gas. The sights are set high, the rear sight forming part of the permanently attached carrying handle. The rifle is laid out in a straight line, instead of the butt dropping as traditionally.

Right: The Russian AK-47 automatic rifle is gas operated; a piston above the barrel being driven back to impart rearward motion to a bolt carrier. This rotates and unlocks the bolt and then draws it to the rear, cocking the firing hammer simultaneously.

Sub-machine guns

The rifle is the most typical and useful personal weapon of an infantryman, but for special situations he may use a *pistol* or *sub-machine gun*. These are both comparatively inaccurate weapons, firing low velocity rounds, so they are used for work at close range.

Behind their personal weapons is a long line of increasingly heavier weaponry to support infantry in battle. The smallest unit of riflemen is normally a section, which contains about ten men and would usually be equipped with a *machine gun* and a one-man *anti-tank weapon*.

Most armies use a general-purpose machine gun (GPMG) which can perform all the necessary tasks. Fairly typical of a GPMG is the British Army's L7A2. While it performs all its tasks adequately it is not entirely satisfactory in any of them. As a light machine gun, which is what an infantry section requires, it suffers from the handicap of weighing more than 12kg (28lb) when loaded with a belt of 30 rounds. That is a lot for an infantryman to run across country with.

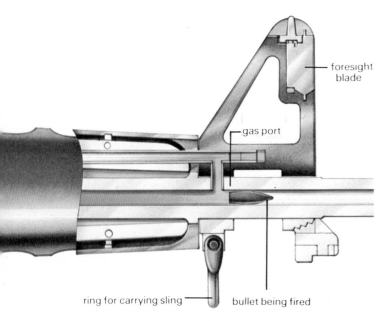

foresight
blade

gas port

ring for carrying sling   bullet being fired

An additional disadvantage is the belt feed, because the trailing edges of the belt can become caught in undergrowth and hedgerow as the gunner doubles to a new firing position. In spite of all this it is difficult to see what sort of weapon can provide a replacement for the GPMG.

Anti-tank weapons

The single man anti-tank weapon which is an infantry section's other support weapon is usually a shoulder-held rocket launcher. Unfortunately for NATO soldiers, their US-designed M72, which launches a 66mm rocket, is not good enough to harm modern tanks unless it is used against the thin armour at the back of the vehicle.

This very serious deficiency in the infantryman's anti-tank capability may soon be ended on the NATO side by a new British light anti-armour weapon (LAW). As far as is known LAW is a throwaway launcher which fires a single highly lethal round. It can destroy any armoured vehicle (including the formidable Soviet T-72 MBT) at ranges up to 300m (1000ft). Infantry faced with an armoured threat do not rely completely on hand-held weapons to combat tanks. The support company's weapons are normally of two sorts—heavier *anti-tank systems* and *mortars*.

Until a few years ago, recoilless rifles and guns of various types requiring a crew of two or three men were the backbone of infantry anti-tank capability, but these are largely being replaced by guided missile systems. However, the recoilless gun has not been completely abandoned and the Soviet 73mm SPG-9 fires a rocket-assisted round which does a very efficient job at well over 1000m (1100yd). The missile systems use a number of methods of guidance, from radio control to wire fed out from the missile to the operator. For the operator himself there are basically two ways of issuing commands to correct the missile's flight. The first is by manual control, using levers or a joystick; and the second is by line-of-sight, in which the operator simply keeps the target in his optical sights for automatic commands of correction to be sent to the missile.

The French-designed MILAN (Missile d'Infanterie Leger Anti-char) is a line-of-sight guided weapon which has become popular with a number of NATO armies; it is relatively portable. Its makers claim that it gives a 98 per cent chance of striking targets 250–2000m (270–2200yd) away—and it hits

hard. The US-made TOW (Tube-launched, optically-tracked, wire-commanded) is easy to operate and battle tested. It is an example of the heavier, vehicle-mounted weapons.

Mortar attack

The mortar, which is often found in the support company alongside these battalion anti-tank weapons, is an exceptionally useful weapon which, in effect, gives the infantryman his own light artillery section. It is basically a steel tube into which a bomb is dropped, to be shot out over a high trajectory arc by an explosive charge. Because of their simplicity, very light mortars can be carried by soldiers at platoon level. These are placed on the ground and hand held, usually for the purpose of providing concealing smoke or illuminating flares in an emergency—but their high-explosive bomb is not to be ignored.

The medium mortars of the support company are more sophisticated and accurate, with an exceptional rate of fire and a very lethal high-explosive bomb. The new British-Canadian ML 81mm L16 is a highly successful example of the type which can throw out 15 rounds a minute to a maximum range of 3200m (2 miles). Just a few of these three-man weapons can provide a fearsome barrage when needed.

The weapons of the support company complete the inventory of a standard infantry battalion as it might be recognized world wide, but not absolutely everywhere—in some countries, the threat of tank warfare in the fast-moving *Blitzkrieg* style have forced infantry to adopt expensive and advanced transport—in particular, the armoured personnel carrier (APC).

Portable missiles

By the early 1970s it had occurred to many soldiers that these sturdy APCs were capable of carrying heavy weapons to support the infantry in assault and defence. At first they simply mounted machine guns by the commander's hatch, but now some extremely powerful weapons are coming into service. The infantry are now equipped with man-portable *anti-aircraft missiles*. The best known of these is undoubtedly the Soviet SA-7 which is a shoulder-fired, heat-seeking missile launcher which has seen wide combat service, particularly with guerilla armies. There are not many shoulder-held, anti-aircraft missile systems available, but the Swedish firm of Bofors produce the RBS 70 which is laser-beam guided. This may well provide a deterrent to any aircraft up to an altitude of 3000m (10,000ft). Certainly the Swedish Army, which demands a very high equipment standard, has put it into service.

But just as infantry weapons improve, so do armour and aircraft defence. As the technological advantage sways from one military arm to the other, the only reasonable certainty is that the basic job of a foot soldier will be unchanging. In the final analysis all his sophisticated support weapons are designed to take him and his personal weapon—his rifle—to victory over similarly equipped enemies: to take or hold ground that is important to his commanders.

Above: The L1A 81mm light mortar in action during a British paratroop training exercise. The mortar is one of the standard weapons used by a battalion support company.

Right: US Marines operating a TOW anti-tank weapon in support of troop-carrying LVTP-7s. The operator keeps the target in his optical sights, the missile being guided by wires.

MIRVS

Any future war between the super-powers is likely to include a series of exchanges at long range, as the United States and the Soviet Union launch space-travelling missiles across the continents. One way of countering an enemy's counter-attack is to deploy a sneaky type of missile—the MIRV.

Initially, missile warfare consisted of launching a ballistic missile into space, and allowing it to drop down again on to one selected target. The enemy's counter-attack was to launch an *anti-ballistic missile* (ABM) to destroy that missile before it could reach its target.

During the 1960s, the numbers and locations of ABMs were limited by treaty, so the Americans decided to concentrate their efforts on fooling the USSR ABM—mainly by providing each missile with more than one warhead, not all of which a single ABM would be able to destroy. All this simple *multiple re-entry vehicle* (MRV) could do was fling its clusters of warheads at a single target. But towards the end of the 1960s, it became possible to direct each warhead independently at different targets—and so in 1970, the US Minuteman III missiles were adapted to carry *multiple, independently-targeted re-entry vehicles*, or MIRVs.

Bus route

The task of aiming the warheads at individual targets is performed by the missile's last stage. It became known as the *bus*, because it carried the warheads as passengers to be dropped off at intervals. Once launched, the bus has enough momentum imparted to it by the booster stages of the missile to reach a target, but it also possesses a guidance system for course correction. The guidance systems used do not rely on ground control, and so cannot easily be jammed.

As the bus coasts through space, it shares its momentum with its deadly cargo. Warheads are ejected usually by a coiled spring, and they race away to their target with the initial speed of the bus vehicle. After each release, the bus will manoeuvre in order to place successive warheads on different trajectories, so ensuring they hit different targets. The sequence of delivery can be varied in a number of ways to confuse enemy defences. Further confusion may be caused by sending two warheads to the same target, but on different trajectories and at different times.

Limited areas

There is, however, a limited area one MIRV group of warheads can cover. This is because the total payload carried by a bus is limited by the power of the booster rocket that puts it into orbit, and this payload has to be shared between the weight of fuel the guidance motors want, the weight of the guidance systems (which depends on their complexity) and the total weight of the warhead load—a measure of the vehicle's effectiveness as a weapon. To allow the greatest weight of warhead, the amount of fuel carried is severely restricted, and this restricts the distance between targets allocated to each bus to a few hundred miles.

All the benefits of separate targeting would be thrown away, however, unless each warhead could be delivered with

Above: A Minuteman III missile being lowered into an underground concrete silo, from which it can be launched, and where it will be protected from enemy attack. The silos can survive a nearby nuclear blast, but not a direct hit.

Left: The three 'dunces caps' on a 'plate' are 200 kilotonne nuclear warheads, capable of enormous devastation. This assembly is of Mk 12 warheads for the missile above; beside them is the missile's protective nose cone.

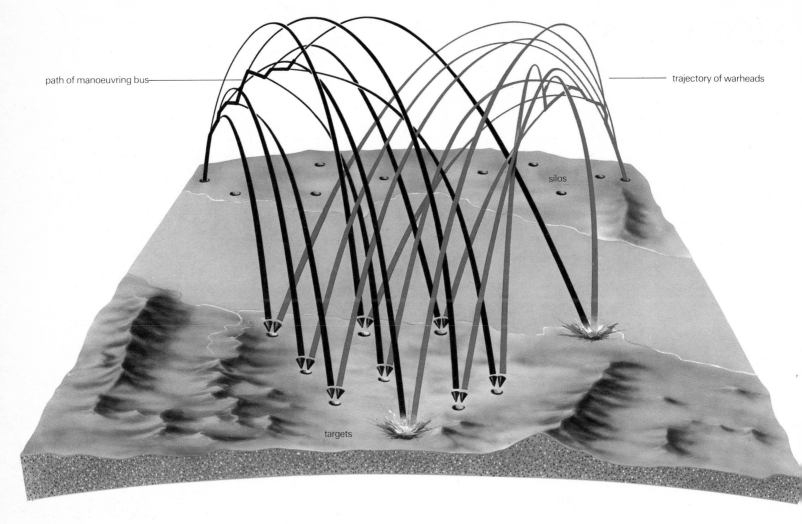

path of manoeuvring bus —————— ————— trajectory of warheads

silos

targets

Above: Just one missile, fitted with MIRVs, can attack many enemy targets; the MIRVs being directed to their targets by the manoeuvring bus vehicle. Fewer launch vehicles are needed (the drawing shows two) and defence is made more difficult.

the necessary accuracy. The accelerometers and gyroscopes that direct long-range ballistic missiles are very precisely engineered, but they are not perfect and, in addition to this, a warhead may be pulled off course by the gravitational effects of the sun and moon or by weather conditions and air density over the target. The result is that US land-launched missiles, for example, have the capability of landing 50 percent of their warheads within 150 metres (495ft) of the target, and submarine-launched missiles can bring 50 percent down within 400m (1320ft) of the target. This is obviously quite accurate enough to destroy cities and similar 'soft' targets, but it is not good enough to cope with 'hardened' military targets such as the reinforced concrete and steel silos in which ballistic missiles are kept ready for launching.

Terrain matching
The problem has been in finding a way to control the flight of the individual warheads. Lately, however, it has proved possible to equip each warhead with an electronic system—

called a terminal guidance system, since it is used at the end of the flight. It is likely that the system uses a technique known as *terrain matching*, in which a flight computer 'looks' at a map of the target area pre-programmed in its memory, and compares it with the ground beneath the MIRV. Of course, a computer cannot use the visual feature of land that a human would do in navigating with a map, but it can make use of a huge amount of other data which various types of electronic sensors can gather—such as change in altitude, sources of radiation and so on.

MX missile
The American answer to the Soviet's MIRV capacity has been as ingenious as its original concept of the MIRV. A new missile system—known as the MX—is under development and should come into operation in 1986. One MX solution to the MIRV threat will consist of distributing a few missiles at random among a very large number of silos. Any large increase in Soviet warheads or in their accuracy can be met with a similar US increase of silos—but empty ones. As the Soviets can never be sure where American missiles are hidden, they would be forced to target every silo in a pre-emptive attack and the Americans can always make sure that there are too many silos for them to do this with any degree of real confidence.

SUBMARINES

Leonardo da Vinci drew one; William Bourne in 1580 designed one; and Cornelius van Drebbel demonstrated one to King James I. But it was not until the early years of the twentieth century that submarines became practicable—at the same time they became necessary forms of modern warfare.

The strength of the main pressure hull determines the depth to which the boat can dive. Made of steel, the hull's plates were originally riveted together, but this has now been replaced by welding. The usual description of 'cigar-shaped' is approximately accurate—the cross section should be circular to obtain maximum strength. A submarine hull should, of course, be pierced by as few holes as possible. But in fact a surprising number are necessary for access hatches, periscopes and so on. Each hole is equipped with a method of closure which is tested to full diving depth pressure.

Outside the pressure hull are the conning tower, the casing and the ballast tanks, which are either of the saddle tank type (great bulges hung from the main structure) or of the double hull type in which a whole skin is wrapped about the pressure hull.

Each ballast tank must have two openings: one at the bottom to let in the water ballast required on diving (and through which that water is expelled by compressed air on surfacing), and one on the top through which air may escape to allow the water to enter. The lower opening can be merely a hole at the bottom of the tank, in which case the tank is known as *free flooding*, or it may have a valve fitted, known as a *Kingston valve*.

The act of diving the submarine is achieved by having all Kingston valves open and then opening the valves at the top of the tank, thus releasing the air pressure which has previously kept the tanks dry. This reduces the submarine to the state known as *neutral buoyancy*; in other words the slightest downward force will cause the craft to sink.

This by itself will not take the submarine below the surface. It must be actively driven below and this is achieved by the thrust of the propellers forcing it forward and rudders directing it downwards. Once below the surface, the craft is adjusted by transferring water in and out of trim tanks so that it lies static and horizontal at the desired depth.

Types of motor

The means of propulsion for submarines was not satisfactorily solved until the diesel engine was incorporated in the design. Petrol (gasoline) engines produced fumes which were occasionally explosive and presented the difficulty of storing a highly volatile fuel. Today, in the normal patrol submarine, diesel engines provide the power required for propulsion on the surface and for charging large storage batteries. As the majority of such boats are now equipped with snort masts, which take air from the surface at periscope depth (up to 15.2m, 50ft), both these functions can now be carried out some way below the surface.

Of course, at below periscope depth, diesel engines can not be used, and so the main propulsion motors are electric, drawing their power from the batteries previously charged by the diesel engines. On the surface or at periscope depth, the main engines can be powered either by direct drive (in which the diesels are coupled to the electric engine through a clutch) or from a generator, again driven by the diesels: either solution preserves the all-important batteries for use at depth.

The need to recharge batteries severely restricts the range and ability of conventional submarines. The only answer is nuclear power. A nuclear submarine needs to refuel—by the

HMS Superb is a nuclear-powered hunter-killer submarine with a crew of 95 men. Nuclear submarines, with their ability to stay submerged for very long periods, are shaped differently from conventional craft. They are the reverse of the normal ship shape, with a blunt rounded forward end, and a sharp run aft. The craft is designed for maximum efficiency under water, and does not perform at its best when surfaced.

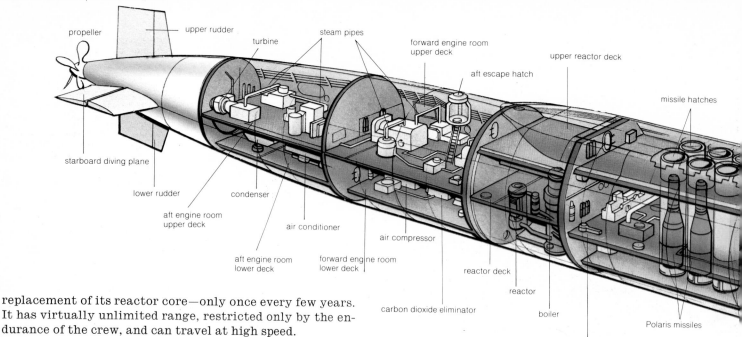

replacement of its reactor core—only once every few years. It has virtually unlimited range, restricted only by the endurance of the crew, and can travel at high speed.

Nuclear subs

This freedom of movement was graphically illustrated in 1958 by the first nuclear submarine, the USS *Nautilus*. Soon after completion, she accomplished the previously impossible feat of running submerged from one side of the Arctic icecap to the other, passing from the Pacific via the North Pole to Iceland.

The nuclear submarine takes its power from one or two pressurized, water-cooled nuclear reactors. The containment of the reactor is surrounded by a jacket, through which water is circulated under pressure. Heated by the nuclear reaction, this water is forced around a closed loop of piping, passing on its way through a heat exchanger where it gives up most of its heat in raising steam. The steam powers the main propulsion turbines and turbo-turbines as well as the submarine's turbo-generators.

The technology is quite conventional: once expanded, the steam condenses and passes back through the heat exchanger. No air is required, and the needs of the crew can be accommodated by chemically purifying (or *scrubbing*) the atmosphere and re-cycling it.

The major threat to a nuclear submarine's cloak of invisibility is its own noise. The anti-submarine war of today pivots largely on whether the hunter or hunted is the quieter. The major sources of radiated noise are the propeller and machinery. Machinery, in particular the pumps of the pressurized cooling system, can be mounted so that their vibration is not transmitted through the hull. Much research is also devoted to designing propellers for silent running, which is more important than their efficiency.

Above and below water

Surfaced, a nuclear boat is out of its element—its whale-like shape is entirely different from the cigar-shape of a conventional submarine. It is also difficult to manoeuvre as the propeller is close to the surface and the upper rudder is almost entirely clear of the water.

Submerged, the nuclear submarine runs silently and easily, but it must operate in a restricted band of water. Should it run too shallow, it will break surface and be detected; too deep and it will be crushed by external pressure. At 30 knots, either limit can be reached within seconds. To avoid serious trouble, a submarine's control surfaces (rudders and hydroplanes) are usually computer-controlled in conjunction with the known response and behaviour of the design.

Prolonged submerged operation would be impossible without the most up-to-date navigation techniques. It is, of course, essential to plot the ship's position. A continuous 'dead reckoning' is maintained by the Ship Inertial Navigation System (SINS) which, by detecting acceleration rates in all three dimensions, can compute a position fairly accurately. Even so, some deviation due to currents may be too subtle to detect, so corrections are made as opportunity permits by satellite or astronomical 'fixes'.

Radio communications to and from a submerged submarine need to be conducted at very low frequency (VLF), because higher frequencies are rapidly attenuated by passage through water and have little range. To avoid the need to surface, the submarine has to stream an aerial, which can be reeled in and out on a winch as required.

Armoury capacity

The submarine now has an overall armoury outranging and outperforming any surface ship except in the sphere of anti-aircraft defence. With the possibility of fitting nuclear heads to any of these weapons, various classes of submarines in the major fleets now carry ballistic missiles such as Polaris (range about 4500km, 2800 miles) or Trident (range about 7000km, 4000 miles) for long range strikes, cruise missiles covering from 400km (250 miles) to short range, and torpedoes equipped with passive or active homing heads as well as wire guidance for close range. Information to control this wide armoury of weapons comes from shore control, satellites, aircraft reconnaissance, the submarine's own radar and sonar, as well as sighting through the periscope.

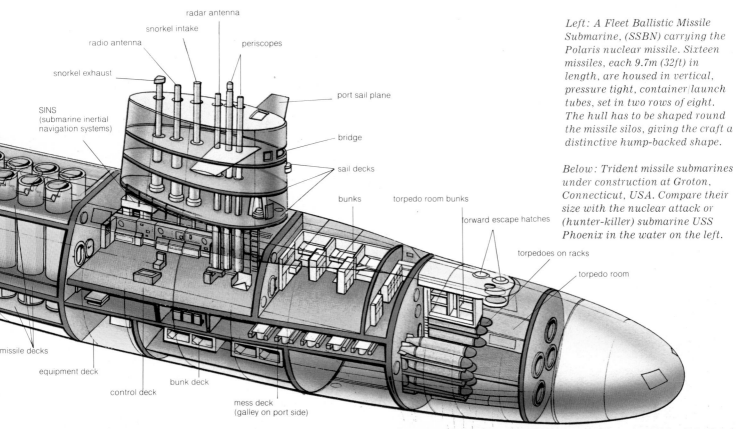

radar antenna

snorkel intake

radio antenna

periscopes

snorkel exhaust

port sail plane

SINS
(submarine inertial
navigation systems)

bridge

sail decks

bunks

torpedo room bunks

forward escape hatches

torpedoes on racks

torpedo room

missile decks

equipment deck

control deck

bunk deck

mess deck
(galley on port side)

Left: A Fleet Ballistic Missile Submarine, (SSBN) carrying the Polaris nuclear missile. Sixteen missiles, each 9.7m (32ft) in length, are housed in vertical, pressure tight, container/launch tubes, set in two rows of eight. The hull has to be shaped round the missile silos, giving the craft a distinctive hump-backed shape.

Below: Trident missile submarines under construction at Groton, Connecticut, USA. Compare their size with the nuclear attack or (hunter-killer) submarine USS Phoenix in the water on the left.

VERTICAL TAKE-OFF PLANE

The year that Man first journeyed into space, 1961, was also the year that saw the most fundamental redesign of the aeroplane. Britain's Hawker P1127 gently lifted vertically into the air, moved slowly forward, then descended like a helicopter. But the plane had fixed wings like a conventional aircraft, and no rotor blades as on a helicopter.

Like most developments in aviation, vertical take-off and landing, or VTOL, became possible only with the design of the right engine. Until the jet (reaction) engine was developed, there was no power plant (apart from a rocket) cap-able of lifting a complete aircraft and an adequate payload vertically from the ground.

Flying bedstead

Between 1945 and 1950, VTOL using jet engines received its share of experiments and tests beginning with the Rolls Royce Nene. Two Nene engines were attached to a frame with four legs, and outriggers supporting small compressed air jets, for stability and attitude control. The device had no wings and supported a single exposed seat from which the pilot would control the aircraft manually. The machine, soon named the 'Flying Bedstead' first 'flew' in 1953. Exhaust channelled downwards through separate ducts from the two engines lifted the 3250kg (7200lb) assembly off the ground. Each engine had a *thrust* of 1800kg (4000lb). The thrust describes the total weight (including itself) an engine can lift:

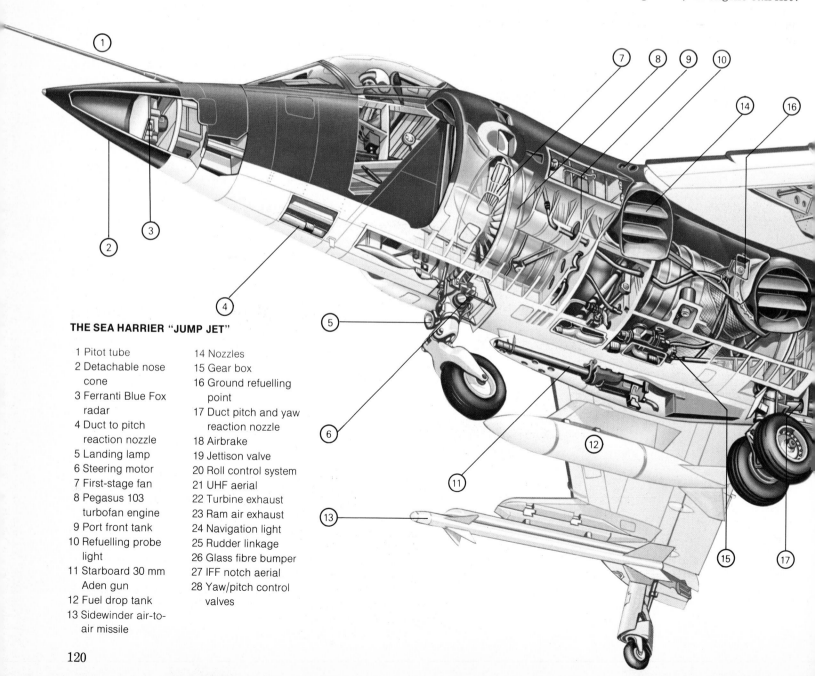

THE SEA HARRIER "JUMP JET"

1 Pitot tube
2 Detachable nose cone
3 Ferranti Blue Fox radar
4 Duct to pitch reaction nozzle
5 Landing lamp
6 Steering motor
7 First-stage fan
8 Pegasus 103 turbofan engine
9 Port front tank
10 Refuelling probe light
11 Starboard 30 mm Aden gun
12 Fuel drop tank
13 Sidewinder air-to-air missile
14 Nozzles
15 Gear box
16 Ground refuelling point
17 Duct pitch and yaw reaction nozzle
18 Airbrake
19 Jettison valve
20 Roll control system
21 UHF aerial
22 Turbine exhaust
23 Ram air exhaust
24 Navigation light
25 Rudder linkage
26 Glass fibre bumper
27 IFF notch aerial
28 Yaw/pitch control valves

as the Bedstead weighed 3250kg, and the total thrust was 3600kg, vertical take-off was just possible.

Rolls Royce quickly took the lead in designing a family of lightweight jet engines built for VTOL aircraft. Prominent among these was the RB 108, which weighed a mere 122kg (270lb) yet delivered a thrust exceeding 1000kg (2200lb). The Short SC-1 delta wing research plane had four RB 108s mounted in pairs firing downwards, and another installed horizontally to propel the aircraft forwards. The SC-1 was the world's first fixed-wing design to achieve vertical take-off and landing, the down-firing engines swivelling backwards a few degrees on their mounting to assist the translation from vertical and horizontal flight, and swivelling forwards slightly to decelerate the aircraft in readiness for vertical descent.

The first vertical ascent was performed in 1958, followed two years later by full transition from take-off to flight to landing. But the Short design did not prove as suitable for operational aircraft as the Hawker P 1127.

Vectored thrust

Instead of five engines in the Short aircraft, the Hawker P 1127 had a single Bristol BE 53 Pegasus with *vectored thrust* through four nozzles, two each side of the engine at front and back. To ascend, the four nozzles pointed down to the ground, gradually rotating to project the thrust behind the aircraft as it translated to horizontal flight. To slow the aircraft down in preparation for a vertical descent, the four nozzles rotated back down towards the ground, then a few degrees forwards to serve as a thrust brake until the aircraft stopped and could be controlled in a slow, powered, descent to the ground.

Unlike the SC-1, in which only 20 percent of the total thrust (one engine in five) could be used for normal flight, all the P 1127s, 5000kg (11,000lb) and more were available for both vertical and horizontal flight. Because nozzle rotation through more than 90 degrees was permitted the aircraft could hover, go forwards or backwards, and move from side to side. Like the SC-1, Hawker's VTOL fighter has gas jets for stability, firing through small nozzles in the wing tips, the nose and the bottom of the tail. Thus balanced on tiny pillars of compressed gas, the stability of the aircraft was assured until transition to horizontal flight allowed the use of conventional lifting surfaces such as the wings and tail.

An enlarged version of the P 1127, the *Harrier*, entered Royal Air Force service in April 1969 powered by Pegasus 101 turbo fans delivering 8600kg (19,000lb) of thrust. Subsequently, they were fitted with the more powerful Pegasus 103 delivering 9750kg (21,500lb), almost twice that of the original engine.

Nautical aspect

One important application of the VTOL concept is nautical, as much as aeronautical. Even the Americans have not got enough of the big, expensive conventional carriers to protect essential shipping from attack by hostile planes, ships and submarines. Joint studies by the US Navy and Grumman Aerospace suggest the value of a smaller ship. This proposed

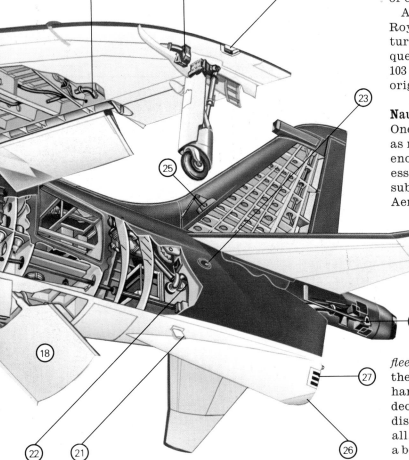

fleet defence carrier would be in the 12,000-tonne class, about the size of a small cruiser, and comprise a large central hangar with a 75m (250ft) foredeck and a 45m (150ft) after-deck. (By contrast, conventional American carriers have a displacement of between 64,000 tonnes and 92,000 tonnes.) In all, the carrier would be approximately 175m (569ft) long with a beam of 20m (68ft) and a flight deck width of 28m (94ft). This Destroyer/Guided Missile/Vertical Take-off aircraft, or DGV, class ship would carry ten Grumman 698 aircraft and up to 120 vertical launch tubes for surface-to-air or surface-to-surface missiles.

The Type 698 aircraft uses another kind of VTOL layout

The Harrier "jump jet" is one of the most versatile V/STOL aircraft. Flying since 1969, it has met or exceeded all specified performance requirements. It is in operation with the Royal Air Force in strike and ground attack/reconnaissance roles, and has been developed as the Sea Harrier for the Royal Navy's Fleet Air Arm since 1980.

adopting a *tilt nacelle* concept in which two engines attached to a common linkage are made to rotate through 90 degrees. Stability and attitude control are secured through vanes placed in the exhaust stream. In other respects the aircraft is conventional, with a fixed wing set high on the fuselage and a horizontal tail on top of the vertical fin.

Vertical take-off may be dramatic, but is not often completely necessary—most of the time, a short take-off run can be allowed. The main benefit from this is that payload capacity is significantly improved. Using a rolling short take-off from the foredeck of a DGV, for example, a Grumman 698 can carry up to 1350kg (3000lb) more equipment than it can with a vertical ascent. An inclined take-off can easily be carried out by planes using vectored thrust—all that is necessary is to pivot the thrust nozzle backwards slightly. Planes with this capability are called V/STOL—for vertical or short take-off and landing. The Royal Navy's Sea Harrier is another example of a V/STOL: this is launched from the so-called 'ski-jump', an inclined plane of 15 degrees from which the Sea Harrier leaps into the air.

Russian model

The Soviet Union's answer to the Sea Harrier is the Yak 36 Forger. This has a single, horizontally mounted, lift/cruise engine discharging through two nozzles placed either side of the fuselage immediately behind the wing. When the nozzles are pointing down the thrust is used for vertical ascent or descent, supplemented by two lift engines fixed to fire down only and mounted in the fuselage immediately forward of the wing. These engines cannot power horizontal flight, which must rely on the single main engine and its two nozzles pivoted back to move the aircraft forward.

Although the single lift/cruise engine is heavier than the Harrier's Pegasus, it generates much less thrust, which of course means that additional lift engines must be hauled along throughout the flight until they are once more brought into use for vertical landing. In contrast, the Harrier's total engine capacity is available all the time. An added disadvantage for the Forger is that failure of one lift engine could bring disaster. Moreover, where three engines are installed the probability of failure of at least one is higher than with a single system operating through all periods of flight. The Soviet aircraft must rely on its lift engines starting up when needed.

The Forger is supersonic in level flight and scores over the Harrier with its computer controlled flight system. Its precision landings compare favourably with the bobbing Harrier, flown down manually through the pilot's control stick, but its overall performance is nowhere near as good, and the Soviets may well be developing a better aircraft.

SPACE

ARIANE

Europe was late entering the space race. Russia launched the world's first satellite in October 1957, followed by the USA in January 1958. The European Space Agency did not send a space satellite launcher into the air until the end of 1979. Yet despite the delay, non-European satellite operators will be more than willing to take a lift on Ariane.

Ariane's main job is to send satellites into space—satellites for scientific research, communications and weather forecasting. During 1979, Ariane I, the first of four test missions before operational flights beginning 1982, was assembled on its fixed launcher pad at the Kourou site in French Guiana. Comprising three separate stages and a fairing over the simulated satellite, the rocket stood 47m (155ft) tall, weighed 211 tonnes and would fire its combined third stage and payload into an elliptical path similar to the track an operational rocket would use to reach a height of 35,750km (22,200 miles).

Geostationary orbit

The height, which if achieved as a circular path is called *geostationary orbit*, is increasingly sought by satellite operators around the world. A geostationary orbit means that the satellite orbits the planet in the same time it takes the Earth to spin on its axis: 24 hours. Because of this the satellite appears to remain fixed over one spot on the surface, but only if the plane of its orbit is in line with the equator. And for that reason, a launch site close to the equator is preferred because it means less rocket power is needed to reach a stationary orbit path.

A rocket launched from Cape Canaveral, Florida, must 'steer' itself into a plane with the equator, far to the south. Ariane, launched from a site much closer to the equator, uses less energy to achieve this orbit and so has a better payload capacity than it would have had from the Cape.

Technically, Ariane is superior to any existing US launch vehicle (with the exception of NASA's Shuttle) comprising three separate stages engineered with more than two decades of experience on rocket motor design. American launch

Ariane's three-stage 211 tonne rocket was successfully launched from Kourou, French Guiana, at 17:14.3 hours GMT on 24 December 1979—early morning, local time. This was a triumph in the project's chequered career.

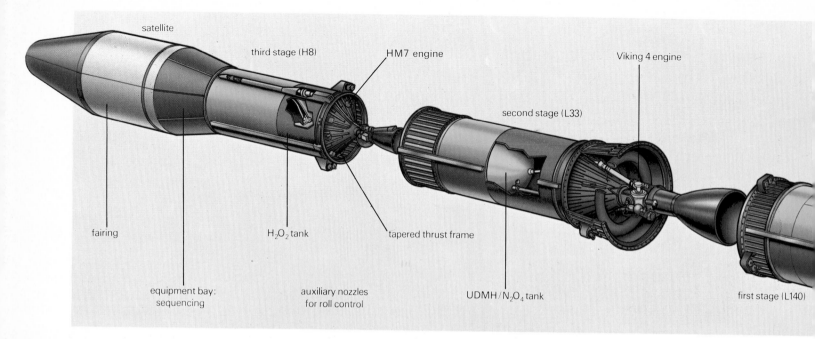

satellite

third stage (H8)

HM7 engine

Viking 4 engine

second stage (L33)

fairing

H_2O_2 tank

tapered thrust frame

UDMH/N_2O_4 tank

first stage (L140)

equipment bay:
sequencing

auxiliary nozzles
for roll control

vehicles all began their development lives before 1960 and, as such, they represent modifications of technology that is now out-dated.

Ariane is able to carry twice as much as the US Delta launcher or almost the payload of an Atlas-Centaur (the largest expendable launcher available for commercial flights). But the development of stretched Delta launcher vehicles, aimed at improving its lifting capacity, stems directly from the competition Ariane is seen to comprise in seizing custom from the United States. So the European Space Agency (ESA) is improving its own launcher and will develop advanced versions of Ariane to keep ahead of the opposition.

Launch procedure

The first two stages of Ariane, called L140 and L33 respectively, burn a form of hydrazine called UDMH (unsymmetrical dimethyl hydrazine) with nitrogen tetroxide as the oxidizer. These propellants are each stored in separate tanks within the two stages. The first stage carries four Viking-5 engines generating a lift-off thrust totalling 245 tonnes. After sending the upper stages and payload to a height of 52km (32 miles), and a speed of 6700km/h (4200mph), the first stage is separated to fall back into the Atlantic Ocean.

The second stage produces a thrust of 72 tonnes from its single Viking-4 engine and propels the vehicle to a height of 144km and a speed of 17,200km/h. Following separation of the second stage, the third and final H8 stage takes over. Burning a combination of hydrogen and oxygen, it generates a thrust of six tonnes and accelerates the payload to a speed of 35,100km/h. When it shuts down the vehicle is coasting in an elliptical path between about 200km (124 miles) at the low point and 35,750km (22,200 miles) at the high point. It is at the high point that the satellite fires its own on-board rocket motor to bring it into a circular orbit at geo-stationary altitude.

The job of the launcher is over when the satellite separates.

Use of high-energy propellants in the third stage marks the technical superiority of Ariane over many contemporary launch vehicles. Only the United States has developed operational rockets using these exotic chemicals and even the Soviet Union has so far failed to get the most from its launchers by using these super-cold fluids.

Improvements

Ariane I will be improved by extending the length of the third stage propellant tanks and by an increase of 9 percent in the thrust of Viking engines employed in the first two stages. In this way, Ariane II will lift 2000kg (about 2 tons) to a stationary orbit transfer ellipse versus the 1700kg carried by Ariane I. Further improvements will be made by attaching two solid propellant strap-on boosters to the sides of the Ariane II first stage. With these, Ariane III will lift 2420kg to the transfer ellipse.

The most developed variant of America's Delta launcher is able to lift 1270 kg. Ariane will, therefore, maintain its capacity for launching two Delta-class payloads. To carry two satellites, a special frame has been developed to sit on top of the third stage and to support, in tandem, separate payloads destined for the same orbit.

Ariane II and III variants are expected to be ready by the end of 1983 while an Ariane IV model is proposed, utilizing four strap-on boosters and a third more propellant in the first stage for a transfer-ellipse payload exceeding 3450kg. As an ultimate 'stretch' of the basic launcher, an Ariane V has been conceived using the first stage of an Ariane IV and much improved second and third stages. With strap-on boosters, this configuration would lift more than 5500kg to stationary orbit altitude and be ready, say its designers, by 1990.

But why, with NASA's Space Shuttle (see pages 136 to 141) already a tested reality, are both European and non-European customers queuing up to use Ariane? Although NASA's shuttle is much cheaper than an expendable rocket used only once, the large weights it is designed to carry mean that the costs of sending a small satellite into orbit by Ariane need be no more expensive than having it flown aboard the Shuttle. Development costs have been much lower, too. Ariane has cost the members of ESA only about one-tenth of the amount America has spent on the Shuttles.

Future orders

Meanwhile, there is much to do in preparing a production run of Ariane I, II and II models to meet an ever-expanding market worldwide. By early in 1981, while preparations were underway for the third test flight, orders for a further 24 launchers had been made known. With an average six flights per year from the Kourou site, Ariane will carry satellites for the prestigious Intelsat organization, a telecommunications consortium handling the international telephone traffic of 105 member states; satellites for the Arab nations to cover Middle East needs; satellites for Europe's own communications and broadcasting needs; and satellites for maritime communications.

Perhaps Ariane's most ambitious launch task will truly set European space technology above other competitors. In 1986, Halley's Comet returns after 76 years for a close pass around the Sun. By 1981 America had cancelled plans to send a space probe to rendezvous with this dramatic visitor but in 1980 ESA formally approved a project called Giotto calling up an Ariane flight in mid-1985 to launch Europe's own space probe to the comet first named by Britain's Astronomer Royal, Edmund Halley, nearly 300 years ago.

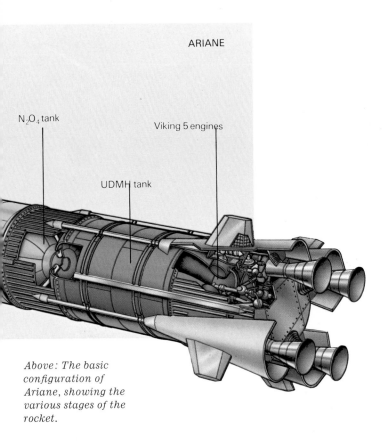

ARIANE

N_2O_4 tank

Viking 5 engines

UDMH tank

Above: The basic configuration of Ariane, showing the various stages of the rocket.

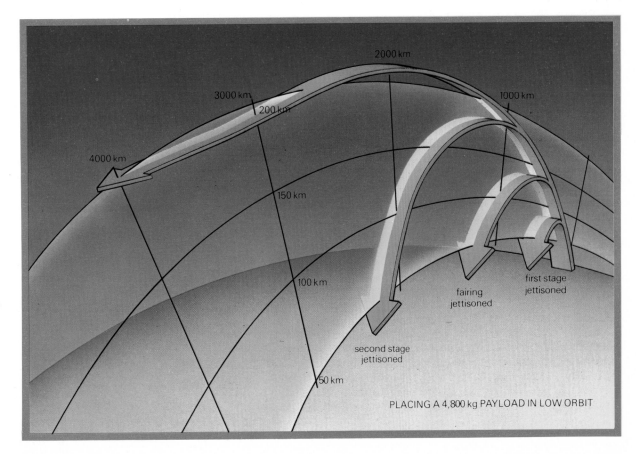

Right: The altitudes and trajectories at which the first and second stages of the rocket, and the satellite fairing, are jettisoned during a mission aimed at putting a 4800kg payload into a low orbit in space.

2000 km

3000 km

200 km

1000 km

4000 km

150 km

100 km

first stage jettisoned

fairing jettisoned

second stage jettisoned

50 km

PLACING A 4,800 kg PAYLOAD IN LOW ORBIT

OPTICAL TELESCOPE

The invention of the optical telescope dates back to the pioneering days of science, when astronomers such as Galileo Galilei first gazed at the heavens in the early 1600s. Through a succession of exciting, record-breaking developments, the modern optical telescope is as important in the Space Age as it was in the Renaissance.

Telescopes collect more light than the human eye, so they can show objects that are too faint to be seen directly. Because they can magnify objects, telescopes show details too small to be visible to the unaided eye. For most modern astronomical purposes, the ability to gather light is more important than magnification. The means by which a telescope collects and focuses light is called its *objective*: the larger the objective, the more light the telescope can collect. A lens, called the eyepiece, magnifies the image formed by the objective.

Refractor telescopes

Some telescopes employ a lens as the objective, and these are called *refractor* telescopes. All early refractors were plagued by optical defects, known as aberrations. Most serious were the fringes of colour seen around an image, caused because a lens brings light of different wavelengths to a different focus. The resulting spread of colours that degrades an image is termed *chromatic aberration*.

Thin lenses with long focal length minimize chromatic aberration, but telescopes that employ such lenses are im-

mensely long and cumbersome, measuring 46m (150ft) or more in length. Suspended by a complex arrangement of poles and pulleys, these 'aerial' telescopes were used to make important discoveries about the solar system, including the rings of Saturn and several of its satellites.

Modern lenses are *achromatic*—producing a virtually colour-free image. Such a lens is composed of two lenses, each to fit a different kind of glass so that the chromatic aberration of one lens cancels out that of the other. The invention of the achromatic lens led to the development of the 0.47m (18.5in) refractor used to discover the tiny, faint, white dwarf star that accompanies Sirius. The two largest refractors ever built—and still in use today—are the 0.91m (36in) of Lick Observatory and the 1.01m (40in) of Yerkes Observatory—both located in America.

There is an upper limit to the size of a refractor, and the Yerkes telescope about marks that limit. The reason is that large lenses tend to sag because they are supported only around the edges. If a lens is made thicker to give it more strength, the glass absorbs some of the light passing through it thereby cancelling out the advantage of a larger aperture. Fortunately a mirror, rather than a lens, can be used as an objective: and mirrors can be supported over their entire back surface. Thus the largest telescopes are all of this *reflector* type. Reflectors do not suffer from chromatic aberration, and are much more compact than refractors.

Reflector telescopes

The light collected by the main mirror of a reflector is focused on to a smaller, flat mirror, which diverts the light into an eyepiece placed at the side of the telescope tube. In

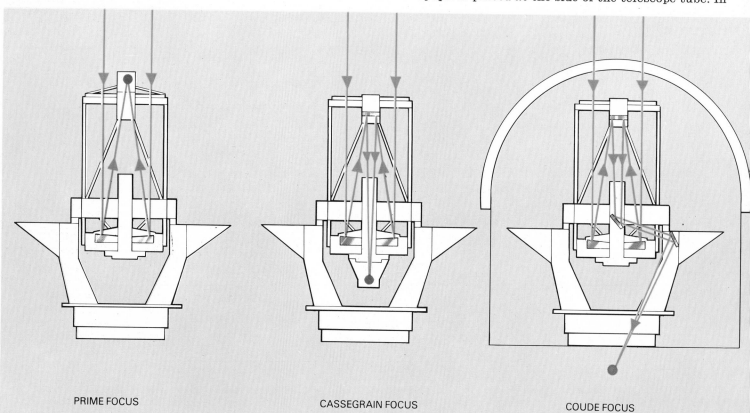

PRIME FOCUS

CASSEGRAIN FOCUS

COUDE FOCUS

The 4.2m (168in) reflector, to be constructed on La Palma Island, has been named the Herschel Telescope after Sir William Herschel (1738-1822) who discovered the planet Uranus. The remote location of the telescope provides favourable conditions for observing faint detail in deep space. The glare of city lights makes it difficult for astronomers at many locations to focus on mere specks of light against a partly lit sky. The Herschel Telescope will also operate in a much cleaner atmosphere because of its distance from major industrial centres. The design embodies three types of focus (below left). Viewing at the prime focus is most difficult to arrange because the observer's cage must be raised by gantry crane to the top of the telescope. The image is unmagnified, but it enables very faint objects to be viewed more clearly than otherwise possible.

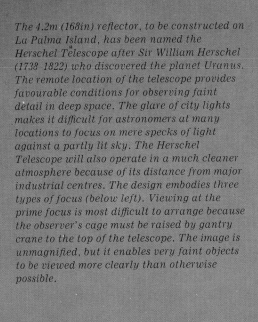

4.2-M LA PALMA TELESCOPE

KEY

1 Observer's cage
2 Pivoted support
3 Secondary mirror
4 Steel framework
5 First Coudé mirror
6 Pivot
7 Mirror cover
8 Main mirror
9 Mirror support
10 Observing floor
11 Turning mechanism

other reflectors, the secondary mirror reflects the light back through a hole in the main mirror, where the eyepiece is placed. The secondary mirror blocks some of the incoming light from reaching the main mirror, but this is a minor disadvantage and does not otherwise affect the image.

Unlike a normal looking glass, in which the back surface is coated with a reflecting layer, a telescope mirror is coated on its front surface. This gives two advantages: the glass does not have to be optically perfect because light does not pass through it; and only one surface of the mirror needs to be shaped and polished.

The first large telescope using a glass mirror was the 1.5m (60in) reflector set up in 1908 on Mount Wilson, California, by George Ellery Hale. George Hale planned a giant 5m (200in) reflector, which eventually opened on Palomar Mountain in California 10 years after his death.

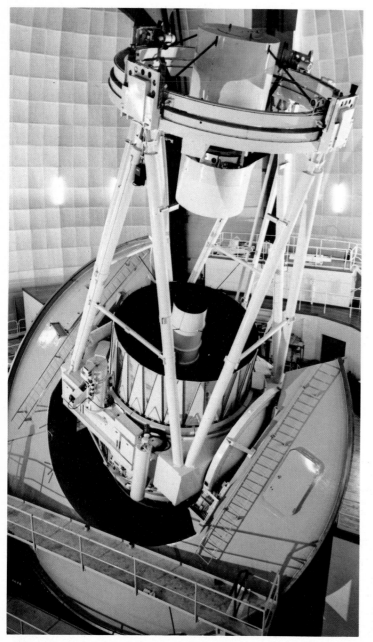

For many years, the 5m Hale reflector was the world's largest and most powerful optical telescope. It can provide magnified images at either a *Cassegrain focus* (at which the secondary mirror reflects light back through the main mirror) or at a *Coudé focus* (which lies adjacent to the main mirror). Conversion from one optical system to the other is simple and quick, making the telescope extremely versatile and adaptable.

Often, even the huge reflector of the Hale telescope might not collect sufficient light to give a clear, magnified image at either of these two foci: a faint image becomes fainter as it is magnified so the design of the Hale reflector enables an observer to sit also near the *principle focus* (which lies opposite the main mirror in the path of the incident light) and observe the unmagnified image. The disadvantage of placing the observer in the light path is greatly outweighed by the advantage of being able to observe a clearer image.

Photographic telescopes

One drawback of conventional telescopes is that they allow only a small area of sky to be seen at a time. To overcome it, in 1930, an Estonian optician, Bernhard Schmidt, invented a new type of photographic telescope that comboned lenses and mirrors to give a wide-angle view of the sky. Today, Schmidt telescopes are an essential part of observatories. Astronomers survey the sky with a Schmidt telescope before homing in on selected objects with their large reflectors.

Contrary to popular opinion, astronomers spend little time actually looking through telescopes. Usually, telescopes are used either as giant telephoto lenses to take long-exposure photographs that can reveal details far too faint to be seen by the eye alone, or they are attached to other instruments, such as *spectroscopes*, to analyse incoming light.

Electronic *image intensifiers* are used to boost the faint light received from the most distant objects; and computers are used to bring out otherwise imperceptible detail—a technique which is already bringing useful results.

The best view of the sky, however, is obtained above the blurring and filtering blanket of the Earth's atmosphere, out in space. Telescopes up to 0.81m (32in) aperture have been flown in unmanned satellites, but in the 1980s, a reflector known as the Space Telescope, with a mirror 2.4m (94in) aperture, will be carried into orbit. The Space Telescope will show details 10 times as fine and objects 100 times as faint as the largest ground-based telescopes. It should answer many questions and give new clues to the origin of the universe. The Space Telescope could produce the greatest advances in astronomy since Galileo turned his puny refractor on the sky.

The 3.9m (153 in) telescope at Siding Spring, New South Wales, is located away from polluted city air—conditions that obscure faint astronomical objects. The optical parts of this reflector telescope are supported by a steel framework which swings in the horse-shoe bearing.

RADIO TELESCOPE

To our light-sensitive eyes, the night sky reveals the planets and the stars in spectacular brilliance, but in scant detail. With the aid of optical telescopes, astronomers have studied astronomical objects in surprising detail—but many objects emit little light, and appear faint and mysterious. If our eyes were sensitive to radio waves instead of light waves, however, the night sky would appear entirely different.

Radio waves and light waves both have the same physical characteristics—they both belong to the family of electromagnetic waves. The main difference between them is in their wavelength. Light has a very short wavelength—less than one-millionth of a metre; radio waves have wavelengths from about one-thousandth of a metre (3mm or about one-eighth of an inch) all the way up to about 10,000m (over six miles). Just as stars emit light, other heavenly bodies can emit radio waves—it is these radio emissions that radio telescopes look for.

To radio sensitive eyes, the sky is dominated by the band of the Milky Way. In fact, the Milky Way is even brighter in the 'radio sky' than the Sun. Stars are generally poor radio emitters, and radio-sensitive eyes would not see any of the familiar stars in the sky at all. There are individual, bright radio sources, though. Some are ring-shaped debris from old star explosions; others woolly looking clouds of gas in space; and yet other sources have a dumb-bell shape, which is typical of exploding galaxies far beyond our own. Such objects emit radio waves but not much light (just as stars emit a lot of light and not much radio emission), and are best studied by radio astronomy.

Receivers first

The first radio telescope was not designed as such: it was just a sensitive radio receiver. In 1931, the Bell Telephone Laboratories employed Karl Jansky, a young engineer, to investigate interference with long-distance telephone calls by 'radio static'. In a potato field in New Jersey, Jansky arranged a large aerial that could rotate on wheels, and pointed it around the horizon to locate sources of static—just as you can find the direction of a radio transmitter by turning a domestic radio aerial until the signal is strongest. As well as identifying static from near-by and distant thunderstorms, Jansky found that static also came from the sky. He had picked up radio waves from the Milky Way.

Dish development

To 'see' the fine details in a radio source, a single aerial is not enough. The first true radio astronomer, Grote Reber (an American amateur radio enthusiast), realized this and constructed the prototype 'big dish' radio telescope. In his backyard, Reber erected a 9.4m (31ft) wide bowl-like reflector, which reflected radio waves from the sky to a focus at the centre of the dish. At this focus, he placed the aerial, and detected radio waves from the region of the sky 'seen' by the telescope. By moving the telescope about, Reber mapped different parts of the sky, using radio waves.

After World War II, radio astronomy attracted large num-

Large-dished, steerable radio telescopes, such as Jodrell Bank in Britain, provide astronomers with extremely detailed radio maps of sources in space, but the clarity and detail of their images have been greatly surpassed by those of the Very Large Array interferometers.

bers of experimenters, who constructed radio telescopes with dishes that were much larger than Reber's. Larger dishes were tried for two reasons. One was simply that a larger dish would collect more radio waves and, therefore, would detect fainter radio sources. The other reason was that a larger dish would be able to 'see' finer details in a radio source: with a larger dish, there is a correspondingly *smaller* region of sky that the telescope can bring to a focus, so, by scanning this smaller 'beam' back and forth, finer structure can be recorded. This ability to see (or resolve) fine detail is a system's resolution or *resolving power*.

The dish design culminated in Sir Bernard Lovell's colossal Jodrell Bank 76m (250ft) radio telescope, the king of radio telescopes for 14 years until the German 100m (328ft) dish at Effelsberg, near Bonn, was completed in 1971. There is an even larger, but immobile, dish at Arecibo in Puerto Rico. This is a 305m (1000ft) hollow in the hills strung with reflecting wires, with the aerial hung above it from immense towers around the dish. Although the Arecibo dish cannot be moved, the telescope's team scans the sky as the Earth rotates.

Interferometers

But astronomers have always wanted to see ever-finer detail in radio sources, and this demands ever-larger telescopes. Engineering problems have restricted the size of radio telescopes, so that Jodrell Bank and Effelsberg are about the upper limit for single steerable dishes. Radio astronomers, however, can detect finer detail by linking two telescopes

electronically. In such an arrangement, structure is seen not by the scanning motion of the telescopes, but by studying the light and dark variations produced when the output voltages of the two telescopes are combined. These variations are called *interference fringes*, and a two-telescope instrument is thus an *interferometer*. The technique has been so refined that radio telescopes on different continents can now be combined. The results of this *very long baseline interferometry* show the detail that would be obtained with a radio telescope the diameter of Earth—'seeing' details a thousand times finer than the best optical telescopes can.

An interferometer can reveal the finest details of a source, but not the broad structure. The pattern of interference fringes for a quasar, for example, might reveal a tiny radio source at the quasar's centre, but not showing the larger regions emitting radio waves around it. Sir Martin Ryle, Britain's Astronomer Royal, has used shorter interferometers in a method that shows moderately fine detail without losing the larger regions—it gives an undistorted 'picture' of a source equivalent to that from a dish telescope several kilometres across. The principle of Ryle's telescopes at Cambridge has inspired the *very large array* in the Netherlands and in New Mexico, where a group of 27 radio dishes are linked in about 27km (16 miles) of desert.

Different wavelengths

Like any radio receiver, a radio telescope must be tuned to a particular wavelength (or frequency) of incoming radiation. Many of the natural radio emitters in the sky emit *all* radio wavelengths, so the radio astronomer can choose which wavelength to receive. But some of the cooler gas atoms or gas molecules emit specific wavelengths: a different wavelength for each type of atom or molecule. So by tuning in, astronomers can identify each molecule, just as a radio listener can tell which station he or she is receiving by noting the wavelength on his tuning dial. Radio astronomers have now identified in space more than 50 different kinds of molecule, ranging from simple types, such as carbon monoxide and hydrogen cyanide (both poisons) to complex, organic ones such as ethyl alcohol.

Left: Interferometry —a technique widely used in physics to make linear measurements—has been applied to radio astronomy with outstandingly successful results. Several telescopes are linked electronically, and their output is combined and processed by a computer which prints a contour map of the detected radio sources.

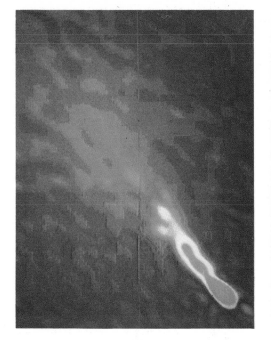

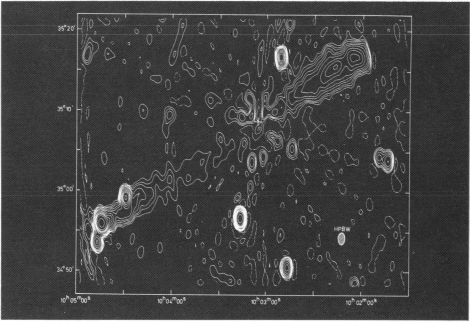

Above: A radio jet streaming from the centre of a radio telescope. An example of the output from a Very Large Array interferometer.

Above right: A radio map showing one of the largest radio sources in the universe. The emissions from the source extend for a distance of about 18 million light years.

In practice, radio astronomers are not free to search the whole spectrum of radio waves. For a start, emissions from space at the same wavelengths as artificial transmissions (such as television and radio) are swamped, because television and radio signals are so much more powerful. To allow astronomers a chance, television and radio broadcasts are forbidden on certain, important frequencies. The second problem is not man-made, but is due to the Earth's atmosphere. This effectively blocks off emissions with wavelengths shorter than about 3cm (1.2in) and longer than about 30m (100ft). This problem could be solved by putting radio telescopes on board satellites.

Radio telescope discoveries
The discoveries that radio astronomers have made, particularly in the past 30 years, range from the near-at-hand, within our solar system, to the most distant reaches of the universe—well beyond the regions that optical telescopes can see.

The Sun's hot, tenuous atmosphere is a radio source; and so are the magnetically disturbed regions of its surface. Visibly, the latter appear only as the dark, indistinguished spots called sunspots but, with a radio telescope, they emerge violently active. During a solar flare (hardly detectable visually) the Sun's radio emission increases by a million times.

Beyond the solar system, the stars are invisible to a radio telescope—except when one erupts in a nova or supernova explosion. The centres of such exploding stars often collapse to become tiny, dense neutron stars, which emit radio waves in pulses—these skeletal stars are *pulsars*. The Milky Way is a strong radio source, but not because of its numerous stars which produce the band of light that can be seen on really dark nights. The radio waves come from the space between the stars, where magnetic fields are deflecting fast-moving electrons—probably shot out from old super-nova explosions.

Beyond our galaxy, radio astronomers have made even more surprising discoveries. All galaxies are made up primarily of stars to which the radio telescope does not respond. But many galaxies also contain hydrogen gas and fast electrons moving through the magnetic fields contained in the galaxy and emitting radiation of all wavelengths. In *spiral galaxies*, all these are highly concentrated in the spiral arm regions, and a radio map often shows up the spiral structure even better than an optical photograph. Spiral galaxies which emit very strongly are called *radio galaxies*. Interferometers show that the radio emission comes not from the region of the galaxy itself, but from two immense 'lobes' of electrons, one on each side of it.

Some extra-galactic radio sources were discovered to be very small in size. Interferometers of the early 1960s could see no detail in them, but the accurate positions indicated that they were the same objects shown as 'stars' on optical photographs—hence they were named 'quasistellar objects', or *quasars*. These are not in fact 'radio stars' but the exploding cores of distant galaxies.

A radio telescope sees a quasar 10,000 million light years away not as it is now, but as it *was* when the radio waves left it 10,000 million years ago. The numbers of very faint, distant radio sources helps us to understand how galaxies changed in the earliest days of the universe—perhaps providing proof for the "Big Bang" theory of its creation.

The many exciting discoveries made with radio astronomy have brought renewed vigour and interest to the study of the universe, but, in answering many of the questions posed by optical astronomy, they have themselves posed new questions for astronomers.

SPACE CAMERAS

Space travel has become so commonplace that the public is now quite blasé about still photographs released when a mission is over. Instead they demand—and get—full-colour live television coverage of all stages of a spacecraft's flight; of astronauts in orbit; and even walking about on the Moon. Yet the problems of 'simple' filming in space or on the Moon are quite large and deserve proper examination.

Initially, space photography was almost an afterthought of space exploration. Only some 300 pictures were taken in the four orbital flights of Project Mercury in 1962. But these photographs became widely distributed, and the authorities began to realize their value—not just for scientific purposes but also because of their immense publicity impact. Thus, in Project Gemini in 1965/66, 2400 frames were exposed in ten flights. And well over 10,000 photographs were taken in just *one* later Apollo mission.

Camera modifications

For the Gemini flights, the basic camera for an astronaut was a Hasselblad. But it was significantly modified, bearing little resemblance to a standard unit. The mirror and reflex viewfinder were removed to comply with National Aeronautics and Space Agency (NASA) regulations limiting the amount of glass permitted on board. In zero gravity conditions, broken glass would float around a capsule and constitute a major hazard. Indeed, NASA had to grant a special dispensation to allow even the glass lenses to remain.

Extra large knobs were fitted to the Hasselblad exposure controls so that they could be operated by a crewman wearing spacesuit gloves. Synthetic 'leatherette' trim was removed from the camera because, surprisingly, it was a potential health hazard. The space modules had a low-pressure oxygen atmosphere, and any materials such as 'leatherette', which would 'outgas' (give off vapours), were forbidden.

The modified cameras had either a non-reflective anodized black trim or a silver anodized trim for lunar surface work. The shiny trim was intended to reflect away heat and prevent the camera from over-heating when outside the spacecraft. Inside the camera, conventional lubricants were replaced by oils that would not ignite in the cabin's oxygen atmosphere.

The standard 'movie' camera used by the astronauts was the Maurer variable-sequence camera. Its main function, however, was not entertainment but data acquisition. Indeed, because of its slow framing rate the Maurer produces a distinctly jerky impression of astronauts' movements when they are played back on television.

Film for space

For space photography, film was modified to give the maximum number of exposures per loading. To achieve this, a special film base of Estar polyester was used which has less than half the thickness of normal triacetate film base. It enables some 200 exposures to be made per magazine loading of black and white film, and around 160 per colour film.

Film was processed under the most stringent conditions by the Photographic Technology Division (PTD) at the Houston Space Centre, USA. Every precaution had to be taken so that invaluable films brought all the way back from the Moon were not damaged during processing. The PTD's elaborate control and back-up facilities ensured success. It is one of the most technologically accomplished film laboratories in the world.

Training for photography

In general, astronauts are not trained photographers. They are chosen for their ability to fly jet planes, or because of their academic training in fields such as engineering or astronautics. While some of the astronauts have been keen amateur photographers, the authorities urge them to regard themselves as 'operators' rather than free-ranging photographers. By encouraging astronauts to follow simple set procedures it was hoped to reduce the chance of error.

Training for scientific or experimental photography is conducted formally, and in simulated mission conditions. General photography is somewhat more informal. Typically, a crew is issued with cameras and other photographic equipment at least six months before the mission takes place so that they can familiarize themselves with them—astronauts due to land on the Moon practised wearing Hasselblad cameras on their chest packs during geological training. Wherever possible, the pictures taken were analysed and problems discussed with NASA photographic specialists. Thus, one of the Skylab crews reported that the training for hand-held photography had concentrated too much on the equipment and its possible malfunctions and not enough on

Right: An astronaut on the Apollo 9 mission stands in the Lunar Module porch, camera in hand.

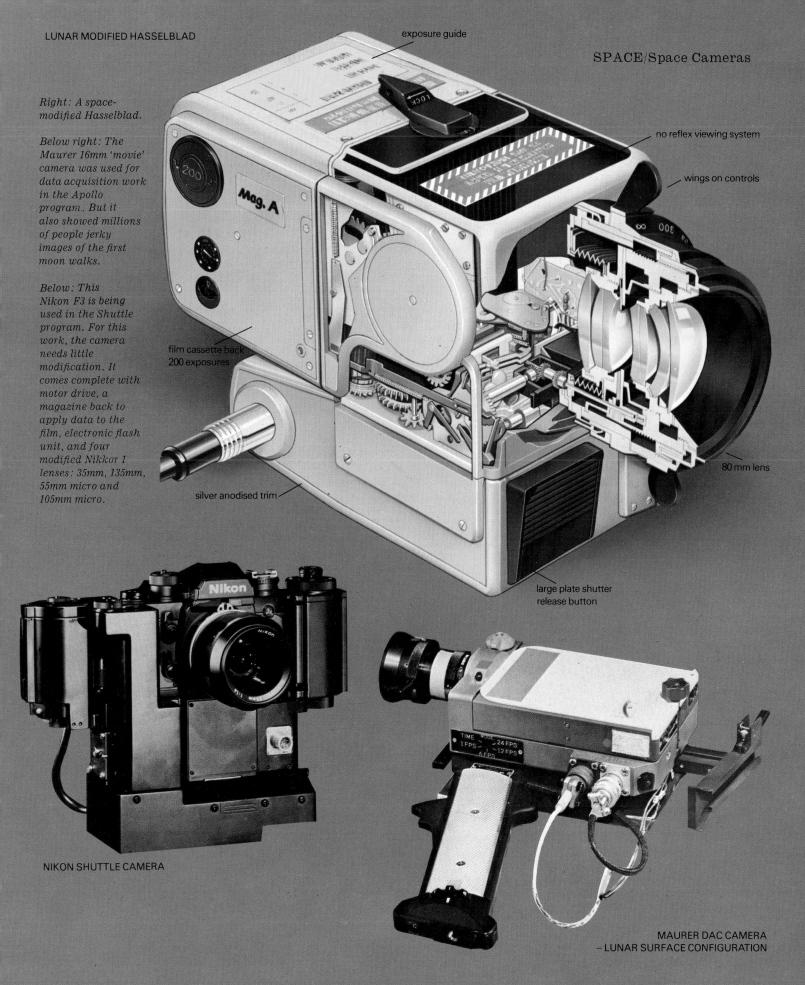

exposure guide

no reflex viewing system

wings on controls

Right: A space-modified Hasselblad.

Below right: The Maurer 16mm 'movie' camera was used for data acquisition work in the Apollo program. But it also showed millions of people jerky images of the first moon walks.

Below: This Nikon F3 is being used in the Shuttle program. For this work, the camera needs little modification. It comes complete with motor drive, a magazine back to apply data to the film, electronic flash unit, and four modified Nikkor 1 lenses: 35mm, 135mm, 55mm micro and 105mm micro.

Mag. A

film cassette back 200 exposures

80 mm lens

silver anodised trim

large plate shutter release button

NIKON SHUTTLE CAMERA

MAURER DAC CAMERA – LUNAR SURFACE CONFIGURATION

133

the end product—that is, the photographs themselves.

Although astronauts face enormous demands on their time and stamina, it has proved unwise to cut down on the time allocated to photographic training. Even astronauts make mistakes, and when this occurs the results can be quite serious. For example, the Apollo 15 crew lost some valuable lunar footage because of lack of familiarity with the equipment. Apollo 8 should have taken pictures of the Moon through various filters on black and white film for photometric analysis. Instead, a colour film magazine was left on the camera. Very slow and very fast black and white films were also confused on that mission and exposed at the wrong shutter speeds. The situation was remedied only by the Photographic Technology Division which fortunately was able to arrange special processing.

In some cases, astronauts have been able to make a significant contribution to the formulation of the photographic program. Astronauts are the operators and are in a unique position to contribute their experience. Thus, during training back in 1969, the Apollo 12 lunar surface crew—Pete Conrad and Al Bean—reported that the shutter speeds chosen for lunar surface photography were too slow to allow satisfactory results. Their recommendations were accepted, and a faster colour film allowing faster shutter speeds was adopted.

Gemini and Apollo veteran, John Young, commander of the first Shuttle test mission, persuaded the authorities to allow him to carry more film than was originally planned. His experience told him that a crew never minded having too much film aboard. But they greatly regretted having to restrict their non-scheduled picture taking. It was often the informal shot taken in a spare moment that resulted in the most dramatic and memorable images.

Space guidelines
Initially, space photographic requirements were stated in general terms and were among the first to be deleted if problems developed during the mission. However, by the time of the Apollo flights, photographic tasks were included in the flight plan.

The Apollo crew found abbreviated but detailed photographic instructions in their flight plan so that the room for operator error was minimized. For example, CM4/EL/80/CEX(f/8, 1/250, focus) 5FR meant that the scene was to be photographed from the Command Module right-hand rendezvous window using the Hasselblad EL camera with the 80mm lens fitted. Ektachrome film would be used and the exposure would be 1/250 of a second at f/8 with the astronauts focusing visually. Five frames of the magazine would be shot.

Exposure help
The most important task for the astronaut photographer is to obtain the correct exposure. But the Hasselblad has no exposure meter, and the peculiar lighting conditions in space would, in any case, be likely to trick any metering device used. A brilliantly illuminated spacecraft against a background of blackest deep space represents just about the most difficult exposure problem there is.

Despite the difficulties, the astronauts had a high success rate—not because of successful metering but because they

Crewmen of Apollo 17 carry out tests on lunar spacesuits on Earth. Hasselblad cameras are mounted on their chest packs.

To frame their subjects, they simply point their cameras in the right direction—no viewfinders are fitted.

used exposure charts. They were built up using the accumulated experience of the early flights. The first missions normally used a fixed exposure time and aperture size 1/125 of a second at f/11, on 64 ASA film. Any changes in exposure had to be related to differences in the brightness of the subject being photographed. Thus, scientists could analyse the images and calculate brightness levels for all types of subject and illumination.

Using this information, the photographic experts compiled exposure charts to guide astronauts on future missions. Each film magazine was given a guidance 'decal' which enabled the men to work out correct exposures without taking meter readings. The lunar guidance panel, for example, takes the form of a clock face. To use it, the astronaut assumes himself to be at the centre with the Sun at six o'clock. The appropriate aperture is selected according to the direction in which the camera is pointed.

Where specific and clearly delineated targets were to be photographed, special meters were used. Called 'spot' meters, these are light meters that record over just a small part of the overall field of view. Unlike normal meters they are not fooled by the bias of peculiar space effects—such as the blackness of space, lunar shadows, Sun glint or light scattered from the Earth's atmosphere. The spot meter is particularly useful for vehicle-to-vehicle photography.

Because the astronauts used Hasselblads that had been stripped of their viewing systems they could not frame their subjects through the viewfinder. Instead, when working on the lunar surface, the camera was attached to the astronaut's chest. To take a picture, all the operator had to do was to point the camera in the right direction. Because the Hasselblad was equipped with a lens that took in a fairly wide angle of view, the astronaut could be fairly sure of cap-

Left: On Apollo 16, commander John Young takes geological pictures on the lunar surface. The lunar module is behind him; a gnomon is in front. Inset: This lunar surface has a clock face display. The astronaut stands in the centre with the Sun at six o'clock. The chart then shows what setting to use, for whatever direction the camera is pointing in.

Below: A gnomon is a tripod which provides geologists on Earth with a chart of reference colours against which the hue of the lunar surface can be measured.

turing the object to be photographed. In special circumstances when it was necessary to frame through a viewfinder, as when using a telephoto (long photo length) lens, sighting devices were available.

Both the solar physics and remote sensing experiments of Earth yielded massive quantities of new and valuable information. More than 40,000 pictures were taken of the Earth, and over 180,000 images of the Sun. Indeed, Skylab pointed the way to a time when one of the major problems was to become that of handling the sheer quantity of images and analysing the data.

Photography problems

Photography in space faces special difficulties arising out of unusual, and often unforeseen, circumstances. Photography seems to have been the last subject in the engineers' minds when designing space craft. Windows, for example, get surprisingly dirty from pollutants such as engine gas, or fog over with internal condensation. Waste liquids dumped out of the space craft tend to follow the craft, glinting in the sunlight and partially obscuring the photographer's view.

On the Moon, surface dust has been a constant menace to photographic equipment. The absence of a lunar atmos-

phere has aggravated this problem. In a vacuum, dust tends to stick to surfaces because of intermolecular attraction. The result has sometimes been a soft-focus effect that gave astronauts the eerie appearance of being haloed in glowing light.

Sometimes, the problems are human. It would have been only natural to suppose that careful plans would have been made to ensure that Neil Armstrong—the first man on the Moon—was subsequently photographed by Aldrin with the Hasselblad. If this had been done, high-quality stills would have been obtained for posterity.

But it did not happen. There are no original still photographs of Armstrong on the lunar surface. The camera never left Armstrong's chest. Though there are photographs of his boots (taken as part of a soil mechanics experiment) and of his reflection in Aldrin's visor, the chance of an epoch-making piece of historical documentation was missed. The fact that Armstrong took so many first-class pictures of Aldrin only underlines the absence of his own portrait out on the Moon.

Despite all of its problems, however, space photography has produced some of the most magnificent images of our era. Through the camera, all of us have gone into space.

SPACE SHUTTLE

Behind the low scrub that binds the sands of Cape Canaveral to the Florida coastline lies a unique runway. A 4.5km (15,000ft) long concrete strip, it is designed to take, not conventional airliners, but the world's first reusable spacecraft —the Space Shuttle.

Space flights—manned or unmanned—are expensive, and no longer is it good enough to plead that money must be spent in the interests of scientific research. More and more, space rockets are used to launch satellites into orbit—satellites that carry out commercial tasks and, therefore, have to prove commercially economic.

Multiple use

One of the reasons for the great expense is that, until now, space rockets were not reusable—they mostly burned away during or after launch. By contrast, most of a Space Shuttle is designed to be used up to 100 times, and could slash the cost of space flight by up to 90 percent. Initial Shuttle plans were for a fully recoverable vehicle, consisting of a winged *orbiter* to carry payloads such as satellites into space, and a system of *rocket boosters* to help launch it. One proposed design was for a 12-engined booster vehicle 70m (230ft) long, carrying a 60m orbiter to a height of 75km (47 miles), at which point the orbiter would propel itself on into space. The payload cost was expected to be very small—around $135 a kilogram, against the $10,000 a kilogram typical of non-reusable launch vehicles. But the maximum payload was only about 11 tonnes, and the development cost would have been a prohibitive $10 billion.

So NASA decided to go for a reusable orbiter with conventional rocket boosters, although that meant the launch cost per kilogram would be two or three times more. The size of the orbiter was reduced, and its carrying capacity increased, by storing its fuel in a disposable external tank. In place of the large manned booster, two strap-on solid fuel rockets were added. Thus the Shuttle evolved to its current configuration.

The orbiter itself is the size of a DC9 jet airliner, 37.2m (122ft) long and with a wingspan of 23.8m (78ft). The astronauts who pilot the Shuttle ride in the nose of the orbiter, as in an ordinary aircraft. Half the orbiter's length is taken up by a cargo bay capable of carrying 29 tonnes into orbit, and of bringing 14 tonnes back to Earth.

At the rear of the orbiter are three large rocket engines; during launch these are fed with fuel from an external tank strapped to the orbiter's belly. This external tank, 47m (154ft) long and 8.4m (27.5ft) in diameter, falls away as the Shuttle reaches orbit, burning up in the atmosphere. Weighing over 30 tonnes, the empty tank is heavier than the orbiter's payload and it is the only part of the system not designed for reuse.

Getting away

To boost the Shuttle at launch, two large solid-fuel rockets are strapped to the sides of the external tank. These strap-on boosters fall away as the Shuttle ascends, parachuting into the sea to be recovered for re-use. Overall, the Shuttle stands 56m (184ft) tall and weighs 2000 tonnes.

Development costs for this design were estimated at around $5.5 billion, but problems and delays eventually increased this figure to $8 billion. One problem was that the engines, which burn liquid hydrogen and liquid oxygen, work

Below and insets right: Stages of a typical Shuttle mission:
1 Two minutes after launch, Shuttle has reached a height of 43km (27 miles) and a speed of 5000km/h (3100mph). The two solid fuel boosters burn out and separation motors push them away. They parachute into the sea and are recovered for re-use.
2 The orbiter's main engines continue to burn until the external fuel tank is exhausted:

this is then jettisoned. Bursts from two small engines nudge the craft into orbit.
3 The cargo bay doors open and a satellite is released into orbit.
4 Another satellite is recovered for return to Earth.
5 The manoeuvring engines fire to reduce speed and the craft begins its descent.
6 The heat shield glows during re-entry as parts reach 1440 degrees C (2900 degrees F).

1 Flight deck
2 Living quarters
3 Steering jets
4 Airlock
5 Remote control arm
6 Temperature control panels
7 Cabin oxygen and nitrogen tanks
8 Cargo bay doors
9 Orbital manoeuvring engines

10 Fuel tank for orbital manoeuvring engine
11 Main engines (to orbit only)
12 Pipes from external fuel tank (jettisoned when empty) to main engines
13 Aerodynamic control surfaces

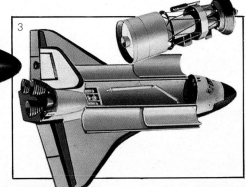

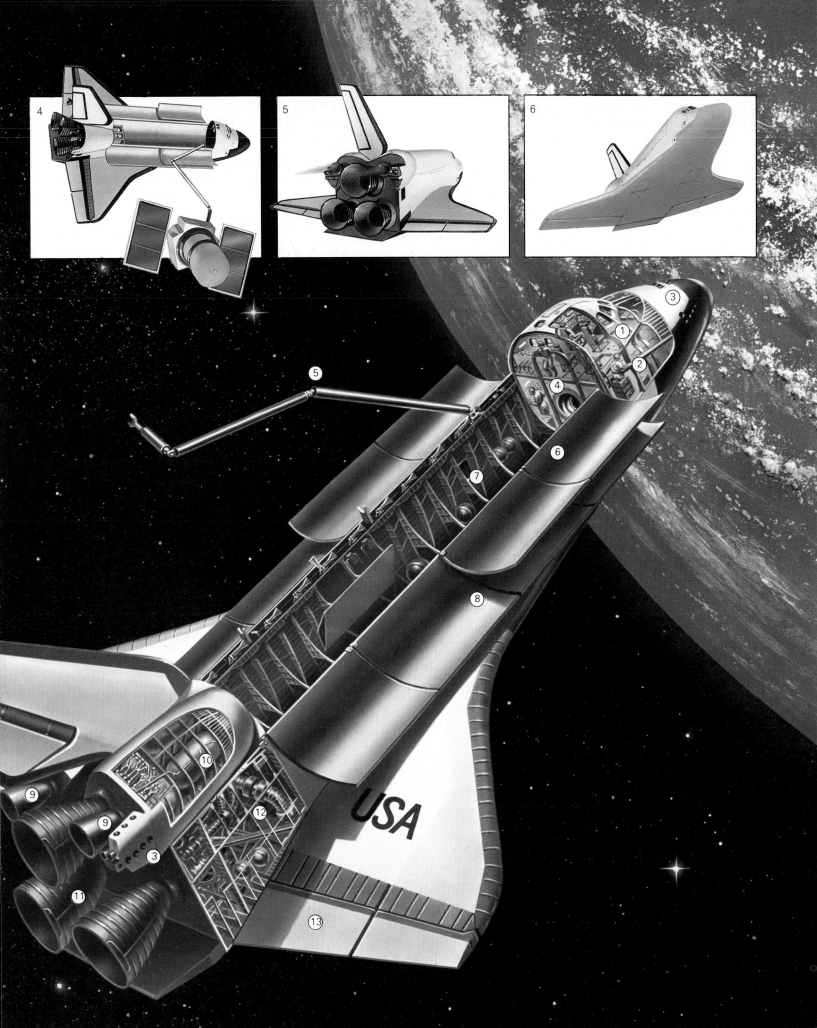

at significantly higher pressures and temperatures than in previous rockets, and of course they have to withstand being fired for up to 100 missions. During the rigorous engine testing program, cracks appeared in fuel pumps and in fuel pipe lines, valves failed and fires broke out. Another problem lay in heat shield tiles that protect the orbiter during its fiery re-entry into the atmosphere. Each of the 31,000 tiles had to be individually shaped and glued to the orbiter's surface: to do this properly took longer than expected.

One vital part of the test program did go successfully: the approach and landing tests, in which a prototype space shuttle orbiter was released from a carrier aircraft over the Mojave Desert, California, and was piloted down to the ground. This was more difficult than it appeared. The orbiter's engines do not operate during the descent, making it less like an aircraft than like a 100-tonne glider. For its atmospheric flight tests, the orbiter, named *Enterprise* after the spaceship commanded by Captain Kirk in the TV series *Star Trek*, rode on the back of a modified Boeing 747 jumbo jet.

The Enterprise was then taken to the Marshall Space Flight Center in Huntsville, Alabama, for vibration tests to confirm that the orbiter's structure would withstand the force of launching. Following these tests, it was ferried to Cape Canaveral where it was mated with a dummy external tank and dummy rocket boosters. On 1 May 1979 Enterprise was rolled out to the launch pad in a rehearsal of launch procedures.

While this was going on, the orbiter scheduled actually to make that first launch, *Columbia*, was undergoing the final stages of construction at Cape Canaveral. Other orbiters in the fleet are named *Challenger*, *Discovery* and *Atlantis*. *Enterprise* itself will never make a trip into space. A plan to bring it up to flight standard was abandoned as too costly. Instead, it was stripped down to provide parts for other orbiters.

The first launch

Launch facilities for the Shuttle at Cape Canaveral have been modified from those built originally for the Apollo program. Shuttles are assembled inside the 160m (525ft) tall vehicle assembly building which once housed Saturn rockets. The completed Shuttle is rolled out on its mobile launch platform by crawler transporters to launch pads 39A and 39B from which men once left for the Moon.

After an initial computer hitch, the very first Space

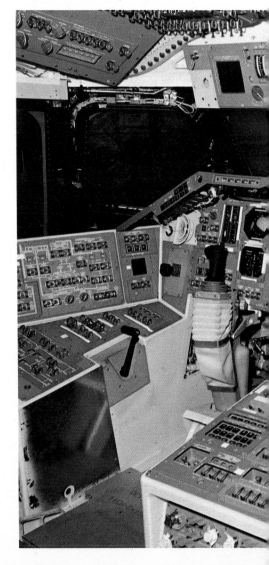

Shuttle flight was planned for 7 am on Sunday 12 April 1981. Right on time, the three liquid propellant engines roared into life and for nearly six seconds burned a transparent flame while computers on the ground examined thousands of separate items aboard the 2000 tonne spacecraft. At the moment of ignition, astronauts Young and Crippen were slammed forward as the nose of Columbia pressed 48cm (19in) toward the big external propellant tank.

Computers then ordered the large solid rocket boosters to fire, which they did—spectacularly. Less than one minute after launch the Shuttle had turned to its proper heading for orbit, gently twisting round its long axis. Two minutes after lift-off the big boosters stopped burning and were jettisoned into the Atlantic, hitting the surface at 96km/h (60mph), their descent rate arrested by three parachutes. The boosters would be recovered and towed ashore for use on up to 20 further flights. Meanwhile, Columbia moved out across the Atlantic, and six minutes later separated from the now empty liquid tank: Columbia was on its own, coasting upward towards a high point of 136km (84 miles) before thrusters fired to push it into a circular orbit 182km (113 miles) above Earth. Back over the United States at the end of the first

orbit, Young and Crippen opened the large cargo bay doors; a vital operation because radiators used for removing excess heat from Columbia were mounted on the inner face of each panel. It could not remain in space for long without deploying the cooling system.

For two days Young and Crippen checked out Columbia's systems, and slept in their flight seats. The cubicles designed into the living quarters below the flight deck were not installed for this maiden flight. On board, the atmosphere was a comfortable oxygen/nitrogen mixture at one-third sea level pressure.

Then it was time to set up the four general-purpose computers for the autopilot re-entry that would make the complex landing operation look so easy. Columbia would be gliding back from space at 27,358km/h (17,000mph), more than three times faster than any previous winged vehicle, building up a temperature of 1482 degrees C (2700 degrees F) on the wing edges and on the nose of the resilient space vehicle.

There was little room for error, but the complex slowing-down manoeuvre and looping glide to landing were achieved perfectly.

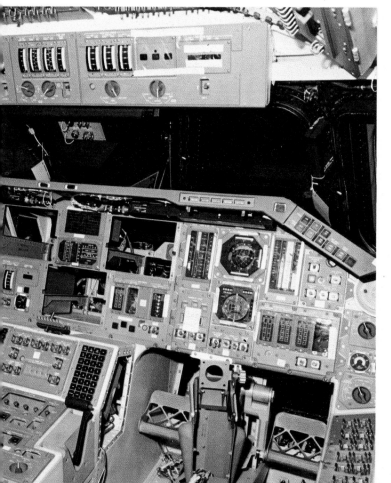

Far left: The whole Shuttle assembly at a rehearsal of launch procedures. The orbiter, on the left of the picture, is dwarfed by the huge tank which provides fuel for the orbiter's engine during launch. This tank is the only part of the Shuttle which is not re-usable. On either side of the fuel tank is a solid fuel rocket motor (only one can be seen in the picture). These provide additional boost during launch; after being jettisoned, they drop into the sea by parachute, and are recovered.

Left: The flight deck of orbiter Columbia nearing completion. It looks more like the flight deck of an aircraft than a spacecraft— as indeed does the rest of this remarkable vehicle.

1 Pilot's controls
2 Spaces for cathode ray tube displays
3 Surface position indicator
4 Warning unit
5 Attitude indicator
6 Vertical velocity indicator
7 Horizontal position indicator
8 Mach meter
9 Cabin air supply
10 Speed brake
11 Computer call-up
12 Fuel cell controls
13 Fire protection
14 Life support
15 Hydraulics panel

Bakersfield

Tehachapi

Mojave Desert

Regular operation

Columbia's first flight was the stuff that spectacular news stories are made of. But it will not be long before Shuttle is in regular operation. Even before the first Shuttle had flown into orbit, it was virtually fully booked for 37 flights, covering three years of operations. The Shuttle can carry several satellites at once in its cargo bay 18.3m (60ft) long by 4.6m (15ft) wide. In orbit, the doors of the cargo bay swing open and an astronaut on flight deck uses a remotely controlled arm to pick out the required satellite and release it into space. The same arm can be used to retrieve satellites for inspection and repair, or simply to bring dead satellites back to Earth, thereby helping clean up the orbital junkyard above our heads. Since the Shuttle has a limited altitude, satellites destined for high orbits must have a small additional rocket stage attached to boost them higher. Space probes destined for the Moon and planets will also have powerful upper stages attached to propel them away from Earth after they have been put into orbit by the Shuttle.

Small, self-contained payloads weighing from 27 to 90kg (60 to 200lb) can ride on the Shuttle cheaply by using one of NASA's so-called 'getaway' specials', at prices ranging from $3000 to $10,000 depending on size and weight. Many users, including private individuals and small businesses, have booked getaway specials.

The most ambitious Shuttle payload of all is the Spacelab space station built for NASA by the European Space Agency —see pages 142 to 145.

Other Shuttles could have their cargo bays modified to carry passengers, making it possible to take a holiday in orbit—brief, expensive, but with sunshine guaranteed.

Columbia's return flight path. At the end of orbit 36, Columbia turned back to front and fired its thrusters for two minutes to slow down and drop out of orbit. Then the Shuttle turned again and pitched nose up so that the protected underside would take the heat of re-entry as it hit the atmosphere 121 km up. Crossing the Californian coast at seven times the speed of sound, Columbia was over 45,000m (150,000ft) above the ground; over Edwards Air Force Base, height was down to 13,500m and speed had dropped to one-seventh. The Shuttle then took a wide looping turn north before returning to the base and landing at just 320km/h (200mph).

1 End of orbit 36
2 Reversal manoeuvre
3 Thrusters fire
4 Re-entry turn
5 Re-entry at 40° pitch
6–9 Positions of Columbia at one minute intervals

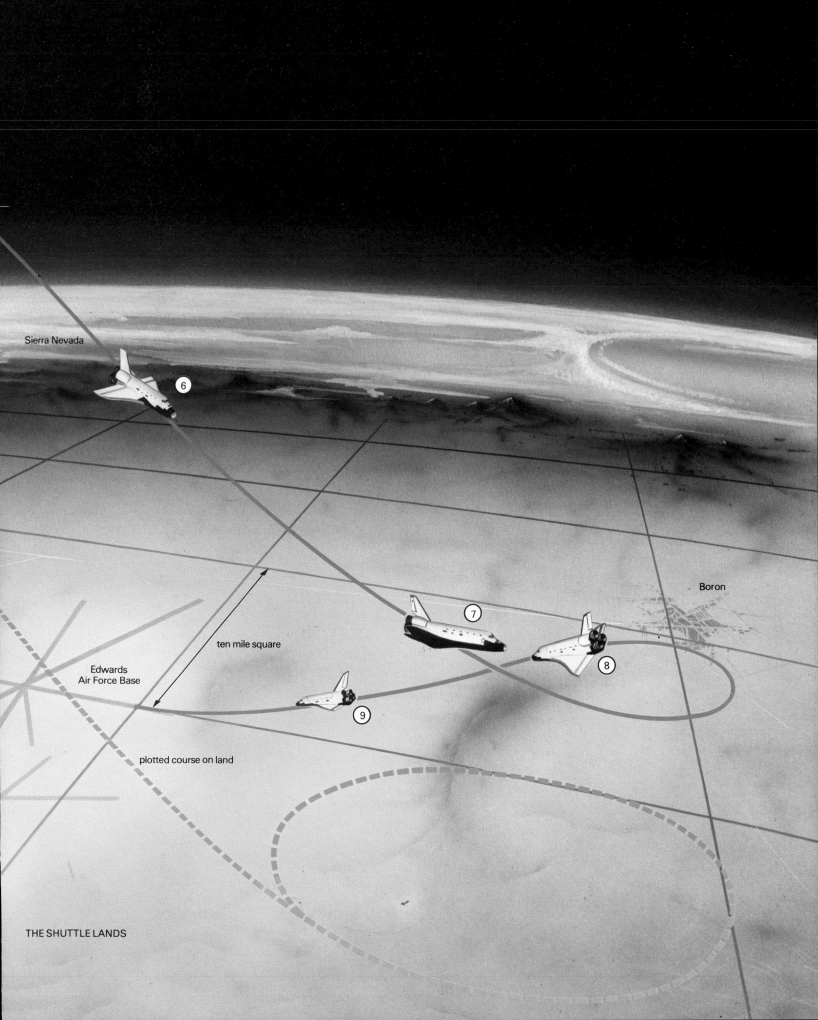

Sierra Nevada

6

Boron

ten mile square

7

Edwards
Air Force Base

8

9

plotted course on land

THE SHUTTLE LANDS

SPACELAB

In orbit above the Earth, the cargo bay doors of the Space Shuttle swing open. Two or more scientists, men and women from the United States and Europe, crawl through a tunnel from the Shuttle's crew compartment into the European-built Spacelab space station, which nestles in the Shuttle's cargo bay. There they carry out several dozen scientific and engineering experiments during a week in orbit. This is a typical space mission of the later 1980s.

Spacelab is built for NASA by the European Space Agency, ESA. The arrangement grew out of a desire by NASA to seek international co-operation, and to share the costs in developing the Shuttle. In 1973 NASA and ESA's forerunner signed an agreement for the European production of Spacelab. Over half the development cost was borne by Germany, with the other European states providing the rest. NASA buys completed Spacelabs as needed to fly in its Shuttle fleet. Unlike previous space stations, Spacelab does not fly freely in orbit, but remains captive in the Shuttle orbiter's cargo bay. More importantly, it can be brought back to earth after each mission.

There are two main parts to Spacelab: a cylindrical pressurized laboratory module in which scientists work, and a series of platforms open to space, called *pallets*, on which instruments are mounted. The pressurized laboratory is 4.1m (13.5ft) in diameter, and can be composed of one or two sections each 2.7m (8.9ft) long, depending on the mission requirements. Each unpressurized pallet is 2.9m (9.5ft) long.

Spacelab can be built up from its component parts in various different ways. It can consist of the pressurized laboratory on its own, laboratory plus pallets, or up to five pallets on their own. The only restriction is that the total mass cannot exceed about 14 tonnes, the maximum that the Shuttle can return with from orbit. This flexibility of design allows Spacelab to be used for a wide range of observations and experiments. When the pallets are being used on their own, the instruments mounted on them can be controlled from an instrument panel on the orbiter's flight deck, or direct from mission base on Earth, although manual adjustments may also be made.

Spacelab can be used for a wide range of experiments,

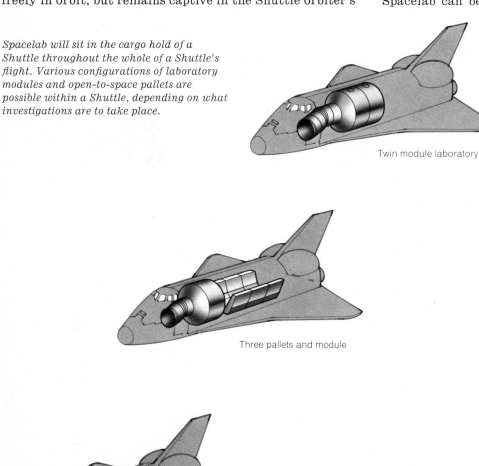

Spacelab will sit in the cargo hold of a Shuttle throughout the whole of a Shuttle's flight. Various configurations of laboratory modules and open-to-space pallets are possible within a Shuttle, depending on what investigations are to take place.

Twin module laboratory

Three pallets and module

All pallets and no lab.

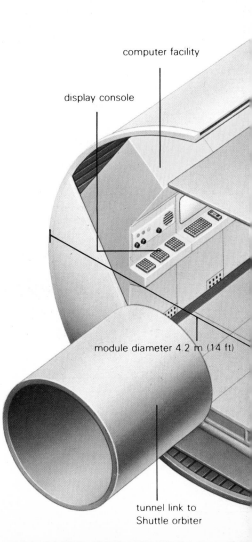

computer facility

display console

module diameter 4.2 m (14 ft)

tunnel link to Shuttle orbiter

Right: A Spacelab consisting of one laboratory module and two pallets. The tunnel allows scientists to pass between the Shuttle's crew quarters and the laboratory; and the airlock allows astronauts to reach the pallets.

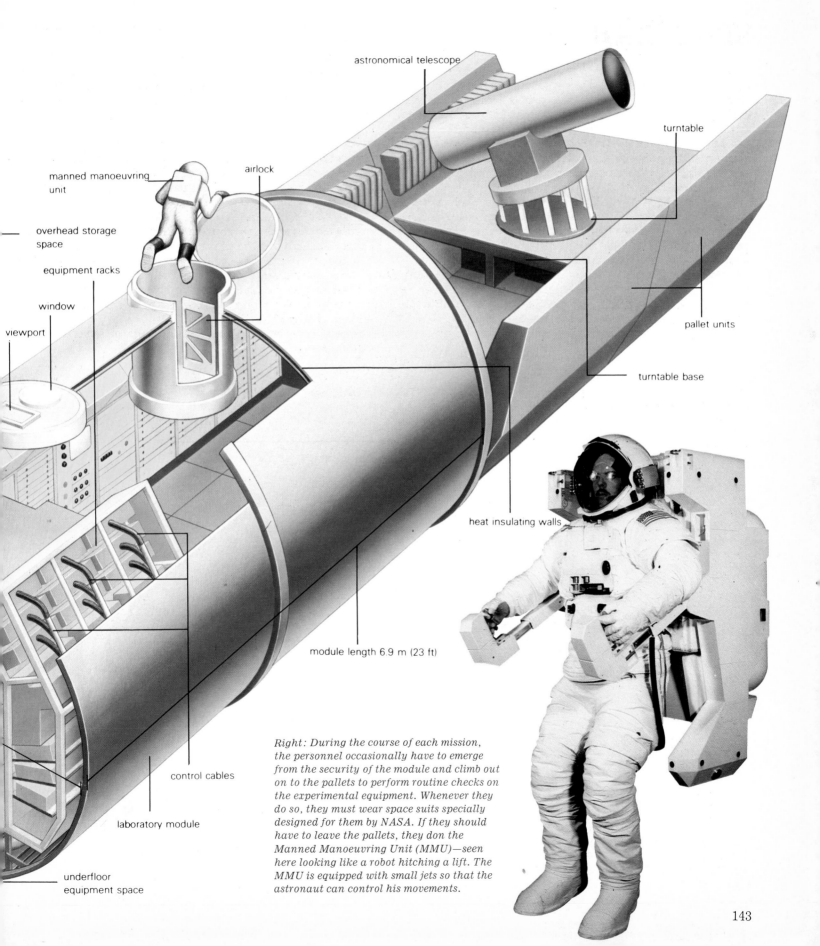

astronomical telescope

turntable

manned manoeuvring unit

airlock

overhead storage space

equipment racks

window

viewport

pallet units

turntable base

heat insulating walls

module length 6.9 m (23 ft)

control cables

laboratory module

underfloor equipment space

Right: During the course of each mission, the personnel occasionally have to emerge from the security of the module and climb out on to the pallets to perform routine checks on the experimental equipment. Whenever they do so, they must wear space suits specially designed for them by NASA. If they should have to leave the pallets, they don the Manned Manoeuvring Unit (MMU)—seen here looking like a robot hitching a lift. The MMU is equipped with small jets so that the astronaut can control his movements.

143

There are no lifeboats in space, so all the equipment must work perfectly. Left in the picture is a full-scale model of the Spacelab module undergoing extensive tests in Germany. The white covering is part of a thermal shroud designed to test the ability of the module's outer skin to keep the lab at constant temperature during missions. In the background (right in the picture) is the real thing: the laboratory module.

ranging from astrophysics, through geophysics and Earth surveying, to biological sciences and processing of materials. For the first Spacelab flight, 76 experiments were selected, 60 from Europe, 15 from the US and one from Japan. Experiments that make use of the conditions of near-perfect vacuum and zero gravity in orbit may form the basis of subsequent orbital industries, once their feasability has been tried and proven.

Scientific help

Qualified scientists must be present on each Spacelab mission to carry out these experiments. Scientists need not necessarily be astronauts—but they must prove themselves to be totally fit in all respects to withstand the medical and psychological stresses of space travel and work. It goes without saying that they must also have excellent scientific and technical credentials. Only after satisfying all these demands will a so-called 'payload specialist' be accepted onto a training course, lasting a year or more.

Then he or she, together with three other payload specialists, a scientist/astronaut and two professional astronaut pilots take off in the Shuttle.

The first task once in orbit is to make sure that all the orbiter's equipment, especially the Spacelab space station in the cargo bay, has come through the launch unscathed. Both the orbiter's crew compartment and the Spacelab working area use a normal Earth atmosphere at sea-level pressure. As soon as the air pressure has been checked, the hatch to the airlock which links the orbiter's mid-deck with Spacelab can be opened, and one or more of the payload specialists can crawl through to complete the equipment checks and adjustments.

A typically varied Spacelab mission might use a pressurized module and two pallets. To the payload specialists floating through the access tunnel into Spacelab's pressure module, everything will look familiar from the many hours spent rehearsing on the ground. Although the lack of gravity means there is no 'up' or 'down' in space, Spacelab has a floor and a ceiling to help the crew keep their bearings while floating freely around. Along the walls are racks of experimental equipment and control panels for the instruments mounted outside on the pallets. Overhead are windows and an equipment airlock through which observations can be made without stepping outside the module.

Space experiments

The equipment outside on the pallets may include a telescope for making observations of the ultra-violet light from stars or, if this is an Earth-pointing mission, the pallets may be equipped with cameras and radar for studying the Earth's atmosphere, land and oceans. The payload specialist's duties may involve aiming telescopes at required targets and making photographic exposures. Other duties might include operating a furnace to make new types of alloy, or crystals of new semiconductor glass for electronics that cannot easily be produced under the gravitational pull of Earth. Other experiments that make the most of the weightless environment of Spacelab may include the processing of super-pure drugs and vaccines, or watching the behaviour of cancerous cells in an attempt to discover reasons for their malignancy.

One major experiment uses the crew as guinea pigs—the 'space sled'. This consists of a seat in which a Spacelab scientist is rocked back and forth along a track to investigate causes of motion sickness in weightlessness. Various astronauts have reported cases of space sickness, akin to car sickness or air sickness, the causes of which are not yet fully understood.

All these are experiments sponsored by university, government or business. But there will be several smaller and

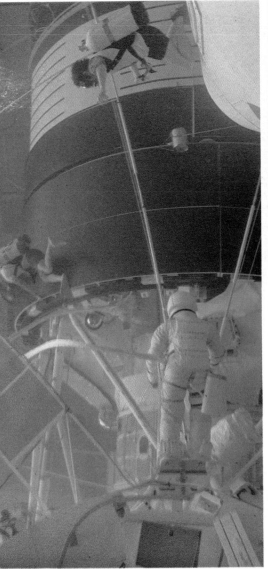

Above: The centrifuge is a terrifying machine that hurls people round at enormous speeds to make sure they can withstand the acceleration experienced during a launch.

Left: Working in weightless conditions is not easy, so training consists of working underwater which simulates some of the problems that will be met in space.

simpler experiments. When NASA's Skylab space station flew in 1973, a number of experiments proposed by high-school students were carried on board, including a famous experiment to see how quickly a spider could learn to spin a web in zero gravity. Following the success of this scheme, young peoples' experiments are also being flown aboard Spacelab missions.

Life in space

Up to three people can work in Spacelab at a time, but often only two are in the pressurized module together. The four payload specialists will work 12-hour shifts in two pairs; time in orbit is limited, so none is wasted. While work goes on in Spacelab, other members of the various research teams are involved in back-up duties on the ground, such as helping sort out occasional troubles with instruments, discussing results to date, and planning new ideas for tomorrow's observing program based on some unexpected new discovery. The scientists on the ground communicate with the payload specialists via mission control in Houston.

Life aboard Spacelab is exhausting but scientifically rewarding. When not at work in Spacelab, the scientists live in the crew quarters situated in the mid-deck of the orbiter. Meals are eaten around a table in the galley. All the food is ready-prepared in cans and bags, some of it needed to be reconstituted by adding hot water. The toilet is as Earth-like as possible, using a flow of air in place of gravity to separate wastes from the body. The crew sleep zipped up in sleeping bags in a curtained-off alcove, and when relaxing, they can take photographs of the Earth, read a book, listen to music on the in-flight stereo or even play a game of computer chess.

In the future

All electric power for Spacelab comes from a series of fuel cells aboard the orbiter. The capacity of fuel cells is limited, so the mission in orbit can last no more than seven days. In future, there will be extra power sources in the form of solar panels which unfold in orbit to provide electricity for Spacelab missions up to a month long. Longer missions will be particularly valuable for biological experiments concerned with matters such as growth rates.

NASA estimates that about 40 per cent of all Shuttle flights will have a Spacelab aboard. In the first few years of Shuttle operations about five Spacelabs a year will fly, rising to as many as one Spacelab a month when the Shuttle is in full operation by 1990. Each Spacelab is designed to last for 50 missions, or a lifetime of 10 years. In the more distant future, Spacelab pressure modules may be combined in orbit to build up large space stations independent of the Shuttle. Such stations could house tens or even hundreds of scientists and other personnel.

At the end of the week-long mission, the payload specialists switch off the equipment aboard Spacelab, and the Shuttle commander closes the doors of the cargo bay securely around it. The crews seal the hatchway leading to Spacelab, and strap themselves into their couches for re-entry into the Earth's atmosphere, in the by-then familiar Shuttle glide style.

VIKING MISSION

In the summer of 1976, two unmanned American Viking spacecraft landed on Mars—the culmination of more than a decade of careful planning. Their mission? To look for life.

Popular interest in the subject of life on Mars is due largely to Percival Lowell, a wealthy American amateur astronomer. At the turn of the century, he observed fine straight lines criss-crossing Mars' red deserts—which he believed to be artificial canals, dug by Martians.

But the canals were shown to be some optical illusion, and it became clear that Mars was too cold, with too thin an atmosphere to support any form of complex life—let alone little green men. But the possibility of simpler life-forms was not ruled out: areas of the surface which change size and shape were thought to be at least areas of moss or lichen. But even these hopes were dashed by the findings of Mariner probes in the mid to late 1960s, as they flew past the planet: there were certainly no canals; the 'moss' turned out to be rocks; and temperatures were even lower than previously thought.

Then in 1971, Mariner 9 went into orbit round Mars to map its entire surface. It showed that earlier probes had, by ill luck, missed some of the most interesting regions. Most excitingly, there was evidence that there had once been water on Mars—a vital ingredient of life. Could any form of life have still clung to existence? The only way to find out would be to land and look: Viking was designed to do just that.

Each Viking consisted of an *orbiter* and a *lander*. The lander, carefully sterilized to prevent contamination of Mars by terrestrial organisms, travelled to Mars attached to the orbiter. The orbiter was designed both to examine Mars from above and to act as a relay between the lander and Earth. Instruments aboard the orbiter included infra-red scanners to detect water vapour and measure surface temperatures, as well as cameras to photograph the surface. An engine aboard the orbiter was fired to put the spacecraft into orbit around Mars.

Viking's experiments

Most of the scientific experiments were crammed into the three-legged lander, which stood 1.9m (over 6ft) tall. The lander was powered by two nuclear generators, so that it was not dependent on sunlight for power as was the orbiter. There were two cameras for photographing the surroundings in both colour and black and white, a seismometer for recording ground tremors, and a boom on which were mounted instruments to measure wind speed, atmospheric pressure and temperature.

Most important of all were the experiments to analyse the soil. Samples could be scooped up by a 3m (10ft) extendable arm. Instruments on board analysed the composition of the soil, looked for organic molecules, and incubated the soil in three different ways in an attempt to encourage the growth of any Martian micro-organisms. All this was overseen by a computer whose memory could be updated from Earth.

Within hours of touchdown on 20 July 1976, the first encouraging news came from an analysis of the Martian atmosphere. In addition to the 95 per cent carbon dioxide already known to exist, there was found to be 2.5 per cent nitrogen,

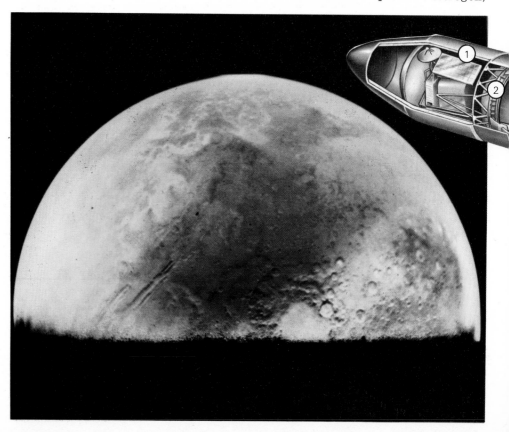

Above right: The Titan III/Centaur rocket slowly raises the Viking I probe at the start of its year-long, 815 million km (506 million miles) journey to Mars.

Far right: The anatomy of the 48m (52yd) tall space rocket.

Right: Pictures taken by Viking Orbiter show Mars to be a red, crater-pocketed planet.

as well as traces of argon, oxygen, carbon monoxide and other gases. This discovery of nitrogen, added to the existence of carbon and water, meant that all three main ingredients for life had now been detected on the planet. But any life which did develop would have to be extremely hardy. For Viking's meteorology instruments showed severe temperature variations: a maximum air temperature of -29 degrees C (-84 degrees F) in mid-afternoon, falling to a low of -85 degrees C (-185 degrees F) at dawn—and this on one summer's day!

Soil analysis

Maximum interest naturally centred on the biology investigations. To conduct sensitive biological experiments automatically on a planet 340 million km (211 million miles) from Earth is a formidable task. The two Viking landers performed magnificently, but the answers they provided are less than clear-cut.

The *gas-exchange experiment* provided results which were typically difficult to comprehend. A sample of Mars soil was mixed with a rich nutrient broth, colloquially termed 'chicken soup', to see whether any changes in the composition of gases above the soil might result and indicate the existence of Martian micro-organisms. Response to the experiment was immediate and dramatic at both Viking sites. As the air in the test chamber was dampened, copious amounts of oxygen and carbon dioxide gushed from the soil sample. When more nutrient was added to soak the soil, oxygen levels dropped but carbon dioxide continued to be given off. What was happening?

Scientists attributed the results to chemical reactions between the soil and the liquid nutrient. The soil of Mars is very dry and highly oxidized (a great deal of oxygen is chemically bound up in it), which accounts for the energetic chemical reactions observed in the Viking experiments. There was no suggestion that the gas-exchange experiment

1 Viking probe
  launch shroud
2 Truss adaptor
3 Centaur
4 Interstage adaptor
5 Stage II
6 Stage 0 solid fuel
  launch rockets
7 Titan III
8 Stage I

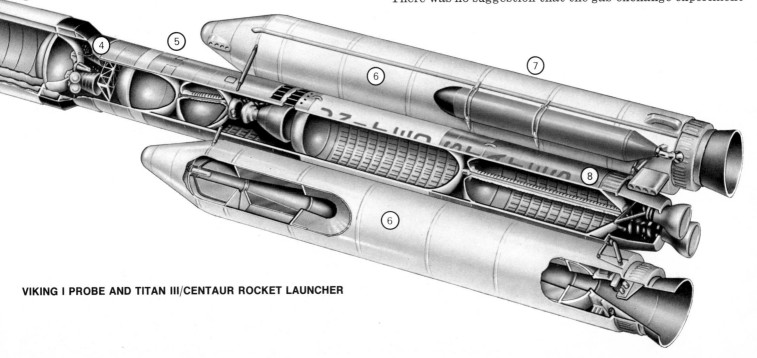

VIKING I PROBE AND TITAN III/CENTAUR ROCKET LAUNCHER

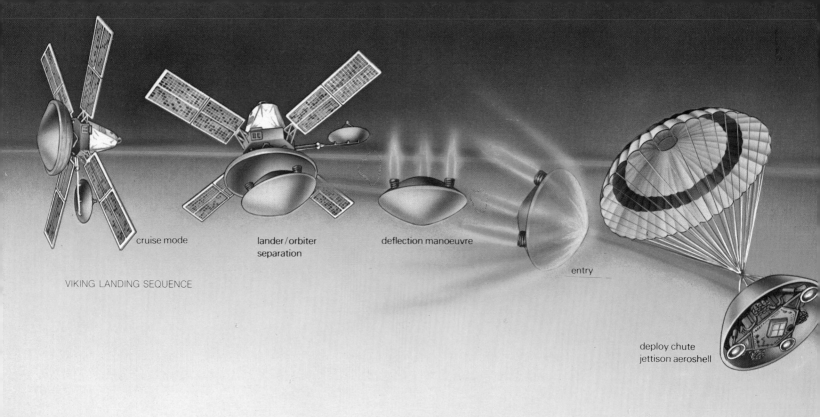

VIKING LANDING SEQUENCE

cruise mode

lander/orbiter separation

deflection manoeuvre

entry

deploy chute jettison aeroshell

had detected Martian organisms, but it provided important clues to the chemical nature of the Martian soil and aided understanding of results from the next experiment, the *labelled-release experiment*.

In this experiment, a Martian soil sample was fed with a solution of nutrients containing radioactively labelled carbon. Any organisms in the soil might feed off the nutrient and release carbon dioxide gas, which would be detected by its radioactivity. As with the gas-exchange experiment, response to the nutrient was immediate. Radioactive gas presumed to be carbon dioxide—the instrument being sensitive only to radioactive carbon—was emitted. A second sample was then heated to 160 degrees C (320 degrees F) to kill any organisms it might contain. When nutrient was injected into this heat-sterilized sample, there was no emission of gas, suggesting that the organisms present had been killed by the heat treatment. This looked good for the existence of Martian life, though chemical reactions were by no means ruled out. As a further test a third sample was given a long incubation period of 60 Martian days to settle the matter. If the gas emissions continued at an increasing rate, that would argue for the growth and reproduction of microorganisms. Alas, the hoped-for increase failed to materialize; at the end of the long incubation period the amount of radioactive gas in the test chamber had dropped to a steady level.

The third biology experiment was known as *pyrolitic release*. This looked for the uptake of carbon from the atmosphere, as would be expected from the action of photosynthesis by plant-like organisms in the Martian soil. In the experiment, a soil sample was incubated under a sample of genuine Martian atmosphere to which had been added some radioactively labelled carbon dioxide and carbon monoxide. A lamp illuminated the test chamber with simulated sunlight, and water could be added if required. The soil was then heated (pyrolysed) to drive off any gases taken up from the atmosphere and these gases were analysed by a radiation

Above and right: Once in orbit around Mars, Viking locates a suitable landing site (inset). The orbiter and lander separate, and the lander aligns itself for

descent. Descent velocity is controlled by parachutes and engines. Sensors on the footpads turn off engines automatically once the craft has landed.

detector. Since it was designed to operate under conditions closely matching those found naturally on Mars, the pyrolitic-release experiment might have been expected to give the best indication of the existence of Martian biology.

Runs of this experiment at both sites showed that something in the soil did seem to fix carbon from the atmosphere; the response was not as strong as from terrestrial soils, but it was positive nonetheless. Were the results due to life? The only way to find out was to heat sterilize the soil and run the experiment again. One sample was heated to 90 degrees C (194 degrees F) for two hours before incubation, which should have been sufficient to kill off any life; but the uptake of carbon by the soil was unaffected. Even after heat sterilization at 175 degrees C (347 degrees F), the reaction was not totally eliminated, confirming that biology could not be the answer.

Inconclusive experiments

All three experiments therefore seem to require chemical, not biological, explanations, although it must be admitted that scientists have not been able to come up with a convincing chemical explanation. A fourth experiment, utilizing a *gas chromatograph mass spectrometer*, also failed to find signs of life.

But all this does not quite rule out the possibility that life may exist in particularly favoured locations, such as around the water-rich polar caps, or that there may be a small population of Martian organisms which did not respond to the Viking experiments. The only way to settle that matter is to bring back a sample of Mars for analysis on Earth.

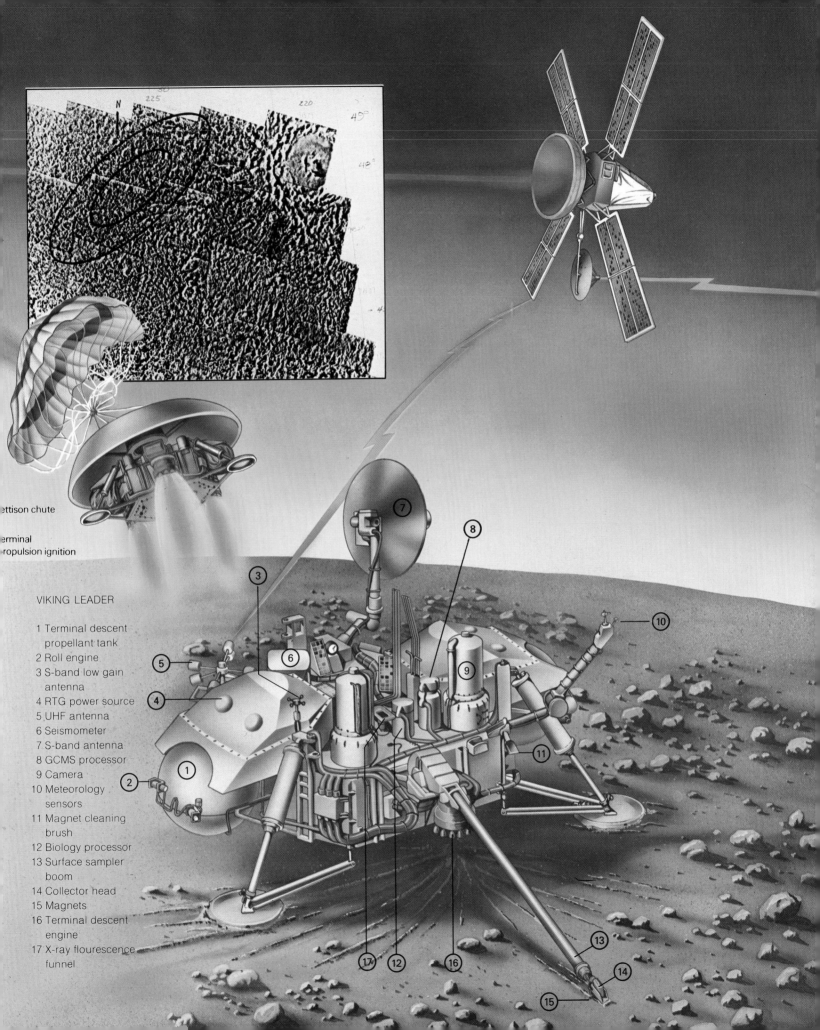

jettison chute

terminal
propulsion ignition

VIKING LEADER

1 Terminal descent
   propellant tank
2 Roll engine
3 S-band low gain
   antenna
4 RTG power source
5 UHF antenna
6 Seismometer
7 S-band antenna
8 GCMS processor
9 Camera
10 Meteorology
   sensors
11 Magnet cleaning
   brush
12 Biology processor
13 Surface sampler
   boom
14 Collector head
15 Magnets
16 Terminal descent
   engine
17 X-ray flourescence
   funnel

WEATHER SATELLITES

To many people, the likely state of tomorrow's weather is merely something of passing interest. To others, such as farmers, builders and other outdoor workers, an accurate weather forecast is essential to allow them to plan their work efficiently and economically. And for those who work on ships or planes, the weather can even be a matter of life and death.

The most crucial stage in the production of a weather forecast is an observation of the weather as it is *now*. To predict weather in the future as complete a knowledge as possible of what the weather is like now must be obtained.

World-wide watch

A constant watch is kept on the weather throughout day and night, on land, at sea and in the air, using a world-wide network of observation stations. These observations are made according to internationally agreed rules and are transmitted around the world as quickly as possible, using a universal code to avoid language problems.

Although some observations are carried out continuously, a large number are made at certain agreed times, especially at midday and midnight Greenwich Mean Time. At these times, the forecasters can correlate all available observations to produce the best possible picture of the Earth's atmosphere at that moment.

A wide range of instruments (from the familiar thermometer which measures temperature at the Earth's surface to more sophisticated devices) is used to take observations on the weather. But one of the greatest advances in recent years has been the use of satellites—out in space, these can monitor weather patterns and movements over large areas of the Earth's surface.

Until recently, the amount of information on weather out at sea was very small. The little there was originated from weather ships owned by various countries and permanently stationed at agreed locations with the sole task of taking measurements of the atmosphere. Data were also sent in by merchant ships using instruments provided by meteorological centres. However, observations were often missing from just those places where forecasters most wanted them, since ships usually try to avoid hurricanes and low-pressure systems with their associated strong winds.

A clearer picture

It was these sort of problems that were overcome to a large extent by the advent of meteorological satellites, which can take regular pictures from above the clouds and supplement observations from ships and remote land areas. They can cover the whole surface of the globe without differentiating between land and sea or mountains and deserts. The result is a much clearer picture of the cloud systems of the world at any one time.

There are two main kinds of meteorological satellite. The first are known as *geostationary satellites*. Stationed about 36,000km (22,400 miles) above the Earth's surface, directly above the equator, they circle the Earth at the same speed at which the Earth rotates. This means that the satellites remain stationary with respect to any point on the surface of the globe, so the pictures they take are always of exactly the same area.

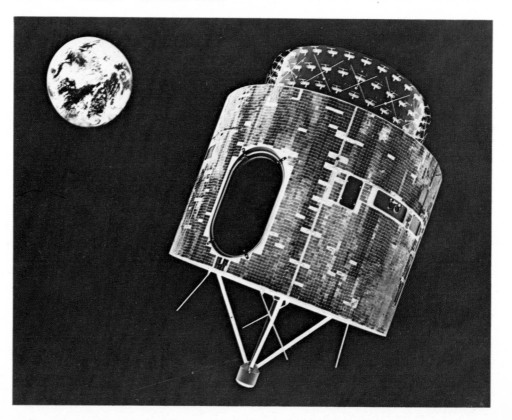

A composite picture of the USA's NASA Synchronous Meteorological Satellite in its geostationary orbit 36,357km above the equator off Brazil.

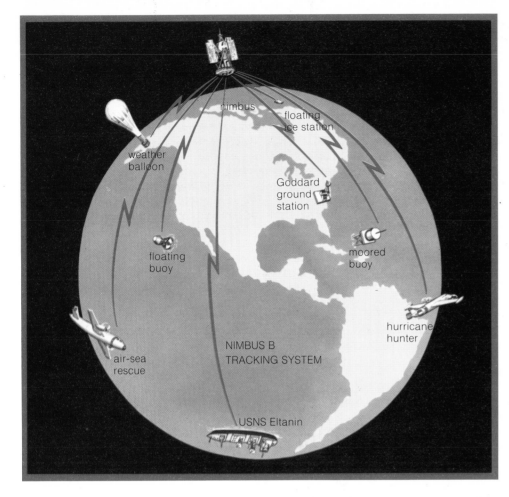

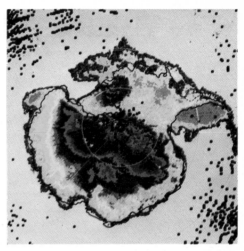

Above: A computer-drawn picture of the southern ice-cap, based on data from the weather satellite Nimbus 5.

Left: NASA's Nimbus B Interrogation Recording and Location System for investigating global weather changes.

A number of geostationary satellites positioned around the equator can effectively cover more of the world. Regular and frequent pictures allow meteorologists to analyse not only the exact positions of cloud systems, including depressions (lows) and their associated warm and cold fronts, but also their precise speed and direction of movement.

Globe trotter

The second type of satellite is known as a *polar-orbiting satellite*. These satellites are put into a much lower orbit than geostationary ones, only about 800km (500 miles) above the Earth's surface. Unlike geostationary satellites, they do not remain over the same point on Earth; instead they circle the globe, crossing the equator every two hours or so and passing close to the North and South Poles in turn before each trip to the equator.

Each orbit is about 30 degrees longitude farther west than the previous one, and the satellite takes a broad swathe of pictures, enabling practically the whole globe to be covered about once every 12 hours. Coverage can be improved still further by launching a pair of satellites 180 degrees apart.

Because they take a relatively long time to produce pictures of the same area, polar-orbiting satellites are not as useful as their geostationary counterparts for assessing the velocity of weather systems. Nevertheless, the high resolution of their pictures, due to their low altitude, make polar-orbiting satellites very useful tools for the weather fore-

caster, particularly those working at high latitudes which are not covered quite so adequately by geostationary satellites positioned over the equator.

Satellite pictures

Weather satellites take two types of pictures. A 'visual' picture, similar to that taken by an ordinary camera, shows the position of clouds in daytime. But the usefulness of such pictures can be greatly enhanced by making an infra-red image, which effectively measures the temperature of the surface at which the camera is directed.

Infra-red pictures, which can be taken in darkness as well as in daylight, show colder surfaces as being whiter. Thus, the whiter a cloud appears on an infra-red image, the colder is its top. As temperature tends to decrease with height, the colder the cloud top, the higher it is in the atmosphere.

Consequently, studying visual and infra-red pictures of the same area side by side can tell the forecaster a great deal. Clouds which appear very bright in both images are probably thick and rain-producing. Clouds which stand out as bright in the visual picture but are barely observable on the infra-red image are probably low and shallow—stratus or perhaps fog clouds.

A combination of infra-red and visual images, together with information from other sources, can also be used to determine wind velocity—a vital parameter for the forecaster. If the forecaster assumes that the cloud moves with

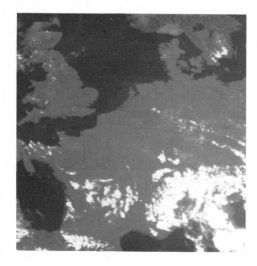

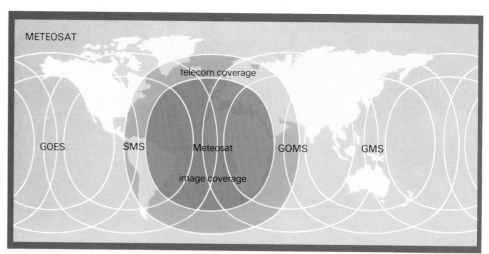

A view of the weather through the eye of Europe's Meteosat 1 (above left). This geostationary satellite forms one of a ring of five, providing complete coverage for low and middle latitudes of both hemispheres (above right) and relaying data to ground stations linked to a computer (right).

the speed of the wind (and the cloud is reasonably thin), then by observing the spread of cloud movement at least an approximation of wind speed can be obtained.

Radiation measurement

Apart from taking pictures, meteorological satellites are also used to measure the microwave radiation emitted by the carbon dioxide which is evenly distributed in the lower atmosphere. The radiation measurements can then be converted into atmospheric temperature. The amount of water vapour in the air can also be determined from satellites using a similar technique. Although neither of these important parameters can be measured accurately or with as much vertical resolution using satellites as with radiosondes (weather balloons), meteorologists are confident that the increased coverage that satellites can give will more than compensate for the lack of resolution.

Equipped with a picture of the weather at a particular time, the forecasters can begin on its analysis, and the actual production of the forecast. This involves making use of large, sophisticated computers to solve the mathematical equations describing the physical state of the lower atmosphere. The work is, unfortunately, not perfect—even with satellites, forecasters do not know exactly what the atmosphere is like at any particular time. Nor are the computer models that represent the atmospheric processes totally complete.

Despite all the technological advances in weather forecasting, therefore, there is still a part for the human forecaster to play in weather prediction. No matter how advanced the computers become, or sophisticated the satellites, the weather forecaster still needs to interpret results, amend them if necessary and, most important, communicate them to people in language they understand.

TRANSPORT

AIRSHIPS

A new generation of airships is rising like a phoenix from the ashes of predecessors such as the R101, Graf Zeppelin, and Hindenburg—names evocative of spectacular and devastating disasters. But with helium replacing hydrogen as the lifting medium, new materials for the airship envelope, and present-day technology, current airships are economic, energy efficient—and safe.

Airships differ from all other forms of aircraft in that they are *lighter than air*: release one from its tethering pole and it will float away of its own accord. By contrast, conventional planes stay firmly on the ground until forcibly given lift—either by the beating of rotors in a helicopter (see page 162) or by the drag of air above and below a carefully-shaped wing (see page 171).

Airship advantages

An LTA craft has several advantages over its HTA (heavier than air) cousins. First, it does not have to waste expensive fuel merely on keeping itself in the air—it needs engines only to propel it forwards, so these can be much smaller than on conventional craft.

An airship travels at low speed, it is true, but even this can be an advantage: coupled with the relatively small power requirements, it means that slow-revving propellers can be used. This cuts noise pollution down to a minimum; indeed, when the craft is operating at its service height, there is no noise.

In addition, the ship's installations and structures are subject to much lower stresses, leading to longer component life and minimal maintenance—freight operators plan to use airships for up to 6000 hours a year. The structure of an airship is simpler than that of a traditional aircraft, too.

Airships are, of course, vertical take-off planes (see page 120), and with none of the problems of needing large amounts of downward thrust to operate in this mode, as 'conventional' VTOL craft do. However, proposed new large airships would have the lift from helium augmented by four turboprop engines, turned so that the propellers pointed upwards, like helicopter rotor blades. Once airborne, the propellers would

Above right: This view of one of the Goodyear airships in flight shows clearly the small passenger gondola with its twin engines. Also visible are the rudders and stabilizing fins that help to control the ship.

Right: On the ground the envelope, or outer bag, looks enormous—as indeed it is. Today's airships are filled with helium—an inert gas—rather than the highly flammable hydrogen previously used.

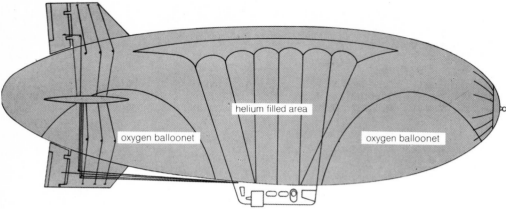

helium filled area

oxygen balloonet

oxygen balloonet

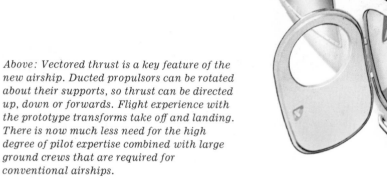

swivel forwards, as on a conventional plane. This *vectored thrust* approach is used also on HTA VTOL planes, but there the transition from vertical to horizontal flight can create control problems: the natural buoyancy of an LTA machine overcomes these transition problems.

The great advantage of a VTOL craft, particularly when combined with relative lack of noise, is that it can pick up and carry passengers and cargo from the point of origin to the point of destination: unlike most other long-distance trips, the entire journey can be carried out without having to change vehicles. This gives the airship a marked advantage over other forms of transport for journeys up to about 500km (310 miles), especially over land and water, and when the journey involves travelling right into the heart of cities. Beyond this distance, the slowness of an airship begins to tell, and using a conventional plane may be quicker.

Most people believe that airships are vulnerable to the weather, but this is not necessarily true. In extensive US Navy trials, an airship survived in weather ranging from tropical storms in the rain forests of Brazil to the severe blizzard and icy conditions off the north east coast of the United States. Despite this, the four Goodyear airships—which form the majority of the ships currently operating—are kept firmly tethered to their masts when there is excessive wind or rain, mainly to ensure that passengers don't experience any discomfort.

Attractive fuels

Yet another advantage of the airship is that, as they already employ huge bags for their helium, there is no problem in equipping them with large amounts of extra storage space. They could, therefore, use cheap, high-volume fuels to drive the engines, such as liquid hydrogen, which is currently not used partly because it is such a bulky fuel. Liquid hydrogen is very attractive as a fuel—it burns cleanly, (its waste product is water) and it is lightweight—see pages 174 and 175. A second alternative fuel, natural gas, is neutrally buoyant —that is, it weighs the same as the surrounding air. This is a distinct advantage for an airship: it means that there is no need to provide a compensation system to alter the craft's buoyancy as the fuel is used up.

At £2.50 per cu m (7p a cu ft) the helium used to lift airships

Above: Vectored thrust is a key feature of the new airship. Ducted propulsors can be rotated about their supports, so thrust can be directed up, down or forwards. Flight experience with the prototype transforms take off and landing. There is now much less need for the high degree of pilot expertise combined with large ground crews that are required for conventional airships.

is expensive—but not prohibitively so in relation to the total capital cost of the craft. Helium is found in geologically old gas and oil fields, and now it is increasingly produced as a by-product of nuclear energy generation. There are large-scale stock piles of the gas in the USA and Poland, so helium scarcity is not likely to be a problem.

Airship designs

There are three main types of airship—*rigid*, *semi-rigid* and *non rigid*. In rigid airships, the shape of the balloon, or envelope, which contains the lifting gas is maintained by a metal framework: the lifting gas is contained in a series of separate cells inside this framework. The Graf Zeppelin and the Hindenburg were examples of rigid airships. Semi-rigid airships have the framework replaced by a single keel running the length of the envelope. In non-rigid ships, the envelope shape is maintained entirely by the pressure of the lifting gas. The Goodyear ships are all non-rigid designs.

Although 'airship' is the modern word for lighter-than-air machines, many other terms have been used in the past, including 'blimp', 'dirigible' and 'zeppelin'. Zeppelin is a term which can be applied accurately only to a rigid airship.

The Goodyear airships consist mainly of the giant cigar-shaped envelope which has a length of 58m (192ft) and a width and height of about 15m (50ft); it is made of rubber-coated polyester fabric. The envelope forms one large compartment without any bulkheads or separate gas bags. Inside the envelope, however, are two air bags, or ballonets, one fore and one aft, which are used to trim the attitude of the craft during flight. The ballonets are inflated during flight by slipstream air from the two propellers which is collected by two airscoops aft of the engines. Fully inflated, the ballonets

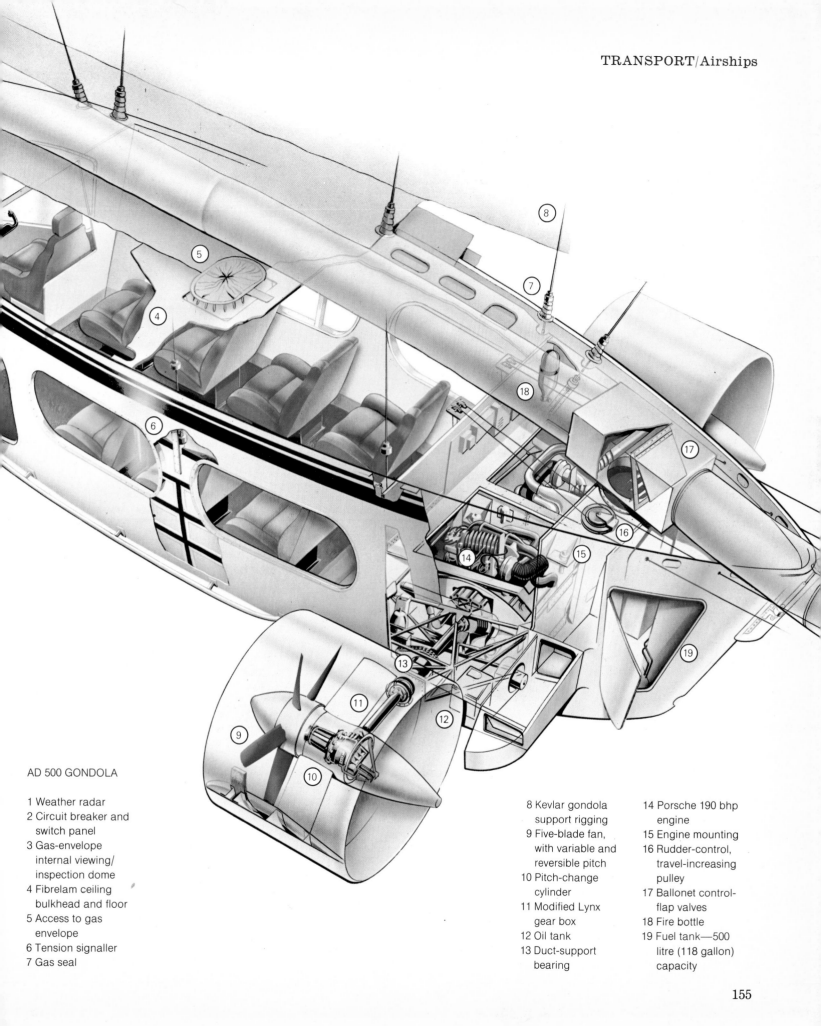

AD 500 GONDOLA

1 Weather radar
2 Circuit breaker and
   switch panel
3 Gas-envelope
   internal viewing/
   inspection dome
4 Fibrelam ceiling
   bulkhead and floor
5 Access to gas
   envelope
6 Tension signaller
7 Gas seal

8 Kevlar gondola
   support rigging
9 Five-blade fan,
   with variable and
   reversible pitch
10 Pitch-change
   cylinder
11 Modified Lynx
   gear box
12 Oil tank
13 Duct-support
   bearing

14 Porsche 190 bhp
   engine
15 Engine mounting
16 Rudder-control,
   travel-increasing
   pulley
17 Ballonet control-
   flap valves
18 Fire bottle
19 Fuel tank—500
   litre (118 gallon)
   capacity

Light aircraft type instrumentation is all that is required in the cockpit of the AD 500. Throttle and pitch are controlled by levers between the pilots.

would occupy about one-fifth of the 5750 cubic metres (202,700-cu ft) volume of the envelope, but normally, they occupy less than one-fifteenth of the total.

By contrast, the passenger compartment or *gondola* is tiny —a mere 7m (23ft) long and 2.4m (8ft) high. The ships can carry just six passengers, plus pilot, and cruise at only 35mph.

Big though the Goodyear ships are, they are tiny compared with some that have been built. The largest non-rigid ship ever was the ZPG3W, a naval airship so big (almost ten times the size of the Goodyear ships) that it could carry a complete radar antenna inside its envelope. However, there is a limit to the size of the non-rigid airships. As size is increased, the relatively low internal pressure becomes inadequate to maintain the envelope shape: a further limitation is that of fixing the rigid payload-carrying structure to the flexible hull. Nevertheless, non-rigid ships are likely to be practicable propositions up to about 122m (400ft) long and with volumes up to 496,000cu m (1,750,000cu ft). Above this size, airships would have to be of the rigid type.

Potential uses

There are many potential uses for airships, including coast-guard duties such as surveillance and pollution monitoring. An airship or plane can search a much greater area of the sea in a day than can a seaship; and airships score over planes because they can travel for much longer without refuelling and, because of their vertical take off and landing capability, do not need an airfield.

At present, passenger-carrying is more for pleasure than business—a task that the airship, with its all-round visi-bility and smooth ride is ideally suited to. Passengers can be carried in a gondola far more spacious and luxurious than an aircraft cabin. But designs suitable for regular city-centre passenger transit are already being developed. Several airships, much larger than those currently flying, are being designed by Airship Industries, Goodyear and other companies. Such ships would be able to carry payloads of up to 80 tonnes and travel at up to 260km/h (160mph); they could be used for transporting heavy cargo, both in military and civil applications, or for moving people quickly and quietly to their destinations.

The design for one Aircraft Industries' ship shows a craft that is 56m long, has a volume of 6000cu m (211,730cu ft) and can cruise with a payload of three tonnes for 58 hours. Propulsive power is provided by two 250hp Porsche engines, driving five-bladed 1.4m ducted propulsors. These are attached to pods towards the rear of the gondola, and provide vectored thrust. Additional forces are generated by control surfaces at the rear of the two vertical and two horizontal stabilizers mounted near the stern of the craft. The combination of vectored thrust, and extensive control surfaces results in a highly manoeuvrable and controllable vessel.

The largest Airship Industries' design is the R150, a rigid airship which is 172m (568ft) long and 153,500cu m (nearly 5½ million cu ft) in volume. The R150 is designed for a parcel delivery firm, who are considering airships as a possible replacement for their fuel-hungry planes; it will be able to carry 80 tonne payloads. With a 58 tonne payload, it will have a range of 4600km. Yet another design is for a short-range airship with a high cruising speed: the TS100 is designed to carry ten tonnes at up to 220km/h.

ELECTRIC VEHICLES

The 'modern' internal combustion engine is an inefficient contraption, using the same basic components that were used in the nineteenth century. Well over 60 per cent of the fuel it burns is converted into wasted heat, which passes into the atmosphere either through the exhaust or via the cooling system. And, of course, it causes pollution. Is an electric car the answer?

In Britain, 45,000 electric vehicles currently do about 300 million miles a year. Many are low-powered milk delivery floats, but an electric van produced by Chrysler UK (now part of the Talbot group), National Carriers Ltd and the Chloride group of battery manufacturers is in assembly-line production. In the US, the postal service uses nearly 400 electric delivery vehicles.

For city driving, one of the electric car's greatest assets is its rapid acceleration from a standing start. It emits no polluting exhaust fumes or noise, and its batteries can be recharged by electricity generated from fuel other than oil (coal, nuclear power and so on). New electronic controls have eliminated the jerkiness with which it once accelerated.

But by mid-1981 at least, three major problems of the electric car are still unsolved. The first is that the lead/acid batteries on which it runs are so bulky and heavy that the car's 'range' between recharging is severely restricted; it cannot carry enough batteries to cover long distances. The second is that its top speed is too low to be acceptable to the average car owner. The third is that, although electricity for recharging the batteries is cheap in some countries, the batteries wear out and are expensive to replace. This means that, per mile, it is far more expensive to run than a petrol-engined car.

A top speed of around 80km/h (50mph) and a range of around 110km (70 miles) is acceptable for short-range delivery vehicles, which can be stabled and recharged overnight. But they are hopelessly inadequate for private cars, which are likely to be called on at any time, sometimes at short notice, and are often used for longer journeys.

Even for a 'city car'—presumably the second car in a two-car family—the minimum targets which General Motors considered acceptable in 1977 were a range of 161km (100 miles) and a top speed of at least 94km/h (55mph), the limit on US roads.

Alternative batteries

So far, several alternatives to the lead/acid battery have been tried. Some of the most effective, such as the sodium/

Right: Delivery vans are about the only vehicles capable of making full use of the present level of battery technology. Batteries usually need recharging every 100km (65 miles) or so—and many delivery vans travel less than this distance in a day. Current batteries are also very heavy and bulky—vans are the only vehicles that have the space and weight to carry them.

Below: The battery pack, fitted into its own cradle, can be removed from the vehicle for recharging.

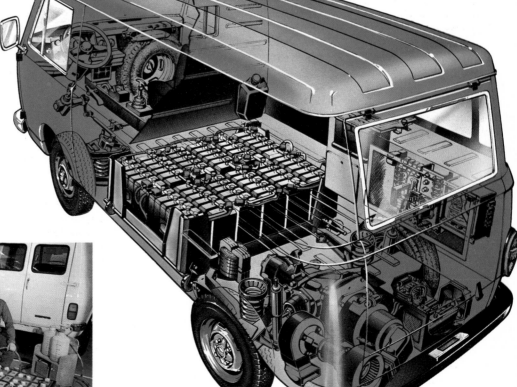

Left: Another possible home for present electric vehicle technology is in the town car. As with delivery vans, the battery's restrictions on top speed and range are not so important—and the quietness and cleanliness of a battery-powered car would be great bonuses in towns.

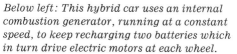

Below left: This hybrid car uses an internal combustion generator, running at a constant speed, to keep recharging two batteries which in turn drive electric motors at each wheel.

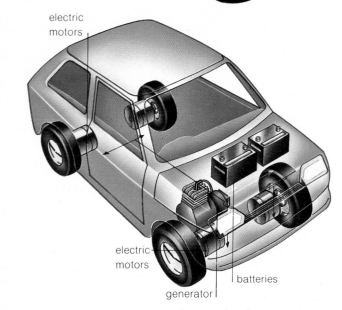

electric motors

electric motors

generator

batteries

sulphur and lithium/iron sulphide varieties, operate at temperatures as high as 650 degrees C—possibly lethal in the event of an accident. Another type, the zinc/nickel oxide battery, promises to store twice or three times as much energy as the lead/acid type, but like other batteries they wear out—and replacements are very expensive.

But while the goal of an ideal car engine battery—low-cost, low-weight, high energy-storage—remains elusive, many researchers have turned their attention to a half-way house: a hybrid car which can offer both petrol and electric motors.

Hybrid cars

Hybrid cars might come in several varieties. In one promising experiment, Ford has adopted a system reminiscent of a diesel-electric locomotive. The car is driven by an electric

motor or motors powered by a heavy-duty battery, and a diesel-driven generator keeps the batteries permanently topped up.

At first sight, hybrid cars look like a waste of time. It is bad enough for a car to have to haul one engine around all the time, but two engines mean extra weight, extra cost, and extra power losses through friction and so on. But the advantage is that the diesel can run at a constant speed—far more fuel-efficient than continually accelerating and decelerating, as in the ordinary car.

In Ford's experimental car, no overnight battery charging is needed. Instead, two batteries provide power for all four wheels, while a turbine generator continuously recharges the batteries. The generator, computer controlled, is said to burn not petrol or diesel fuel necessarily, but 'anything combustible'.

Other hybrid systems have an engine (which could be petrol [gasoline], diesel, Stirling or gas turbine) in parallel with a motor-generator and battery pack.

For short journeys, the vehicle is driven only on the batteries, which can be recharged overnight if necessary. For longer journeys, the engine can be used either to top up the batteries or, when the car is climbing or accelerating, to augment the power of the electric motor.

Another line of development is in *regenerative braking*—as the car is slowed down, the electric motor is used as a generator, feeding 'free' energy back into its own battery pack.

Lightweight cars

The prospects for hybrids are encouraging enough that a number of manufacturers are investigating them. Of course, any electric car of the future will have the benefit not only of advances in battery and motor design, but also the advances in body and transmission design that are already being applied to conventionally powered vehicles. Cars will be built of lightweight materials, and designed to reduce losses by friction as far as possible.

FIRE ENGINES

The Greeks and Romans built the first fire engines—small man-powered water pumps that could be moved about on wheels or skids. Large quantities of water are still the best way of putting out a fire in a burning building—but fire engines have come a long way since those early Greek models.

Man-powered pumps (reinvented in the sixteenth century) were replaced by steam-powered pumps in the 1860s, and these in turn were replaced by petrol-powered pumps from about 1900. These petrol-powered pumps were then placed in lorries and the fire engine (or fire appliance) as we know it was created. Most British fire engines are either *pumps* or *water tenders*.

There are two types of pump, the main difference being whether or not they can carry a ladder escape. Many modern machines are dual-purpose—that is the fittings for the escape are removable so that machines can carry one or not

as the need arises. Pumps have a powerful water pump capable of supplying at least 2200 litres (500 gallons) of water a minute and often over twice as much, and a small water tank holding at least 500 litres (100 gallons). The appliance carries a crew of six—more in rural areas—and has VHF radio to communicate with other engines and their base. Most crews are also equipped with UHF radio for use as walkie-talkies. The pumps have two 20mm (¾ inch) hose reels, one on each side, with their own 110 to 220 litres per minute (25 to 50 gallons per minute) pumps.

Despite their large size and considerable weight, pumps can attain 65km (40mph) in a maximum of 27 seconds from a standing start. Fire brigades have to respond to fires immediately, so pumps have a maximum speed of over 100km/h (60mph). A typical pump, the Albion-Carmichael Firechief, is powered by 125bhp 6.5 litre six-cylinder Leyland diesel engine, with a six-speed manual gearbox driving the rear wheels. The water pumps are driven by the main motor, and the main pump can be mounted in the middle or at the rear of the engine. Pumps designed to carry wheeled escape ladders generally have the pump controls at the side so that the latters' wheels do not get in the way.

Water tenders

In towns, an adequate supply of water is usually near at hand, but where this is not so, fire brigades might use water tenders. These carry a water tank holding 2000 litres (400 gallons) of water. Type A water tenders do not have their own pump, whereas Type B machines have a pump and may also carry an escape ladder. Only Type B water tenders are now being made. In effect, they are pumps with a larger self-contained water supply, a slightly lower performance and less equipment.

Water tenders are not the only way of supplying large

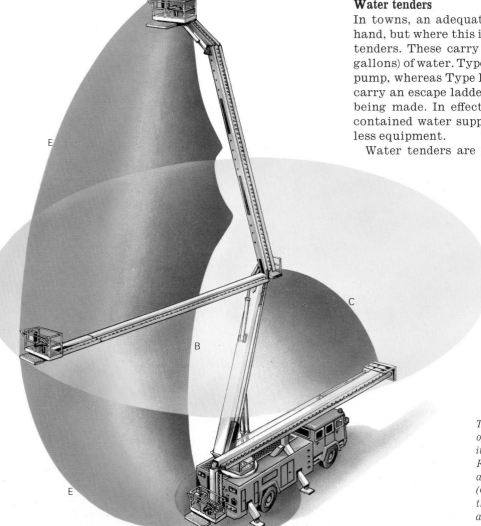

The hydraulically operated articulated booms of this Simon 'Snorkel' fire-fighting unit make it manoeuvrable in several different planes. First (A) the main boom can revolve on its axia; then (B) the platform rises vertically; (C) the extension boom moves diagonally; (D) the platform reaches out through 360 degrees; and finally (E) the platform can move outwards and upwards.

quantities of water; some fire brigades also have hose layers to connect the fire engines to a distant source of water. These hose layers carry nearly 2000m (6000ft) of water hose in two troughs in the back—carefully arranged not to tangle when they are laid out—at spreads of up to 50kmh (30mph). This hose is normally of 90mm ($3\frac{1}{2}$-inch) diameter.

The 16m (50ft) wooden wheeled extension ladders that used to be carried on some Pump escapes have now virtually been discarded. They have been replaced both by lighter metal extension ladders and by special vehicles which can fight fires and rescue people from much greater heights. The older type is the *turntable ladder* fire engine. This carries an hydraulically powered turntable on the back of the engine with a ladder capable of extending to over 30m (100ft). Modern turntable ladders have to be able to be used at any angle up to about 78 degrees, to elevate to 75 degrees in 30 seconds and to extend fully in 35 seconds. The top part of the extending ladder has a fixed hose on it to which a long flexible hose can be fitted before the ladder is extended so that the machine can be used to spray water on the fire from a great height. It can also be fitted with rescue cradles so that several people can be removed from a tall building at once.

The newer type is the '*Snorkel*' or *hydraulic platform*. This has been developed from the sort of vehicle used to mend light standards or perform other maintenance jobs at consider-able heights, and consists of a set of hinged booms with a platform on top. Like a turntable ladder, the fire brigades' hydraulic platforms can be used both to fight fires and to rescue people. The platform can take a load of up to 360kg (800lb) and it can be elevated to maximum height in 80 seconds.

Foam tenders

A completely different type of fire engine is the *foam tender*. Town and country fire brigades have very few of these vehicles, but almost all airfield fire engines are of this sort. Their main purpose is to smother the large oil fires that can occur when aircraft crash. The foam can be made either mechanically or chemically; carbon dioxide gas, which has the same effect, is sometimes also carried. Aircraft often crash well away from the nearest road, so airfield fire engines normally have four- or six-wheel drive and a good cross country performance to reach the aircraft and douse the fire in time to save the passengers. A typical example carriers 900 litres of foam concentrate and 6800 litres of water. Its foam monitor can discharge up to 28,000 litres (6300 gallons) of foam per minute at a range of up to 60m, and this can even be operated on the move. It has two small foam guns and can accelerate up to 80km/h (50mph) in less than 35 seconds. This foam tender is operated by a crew of three, though it can carry more for rescue work.

There are many other types of fire engine. It is not always convenient to have a large machine, and many brigades and firms have small engines based on Land Rovers or similar vehicles. There are also much larger machines. One Ameri-can appliance is the *super pumper complex*. This consists of a vehicle with a pump capable of moving up to 40,000 litres of water per minute, large articulated water tender and up to three satellite tenders. It is designed to cope with large city fires.

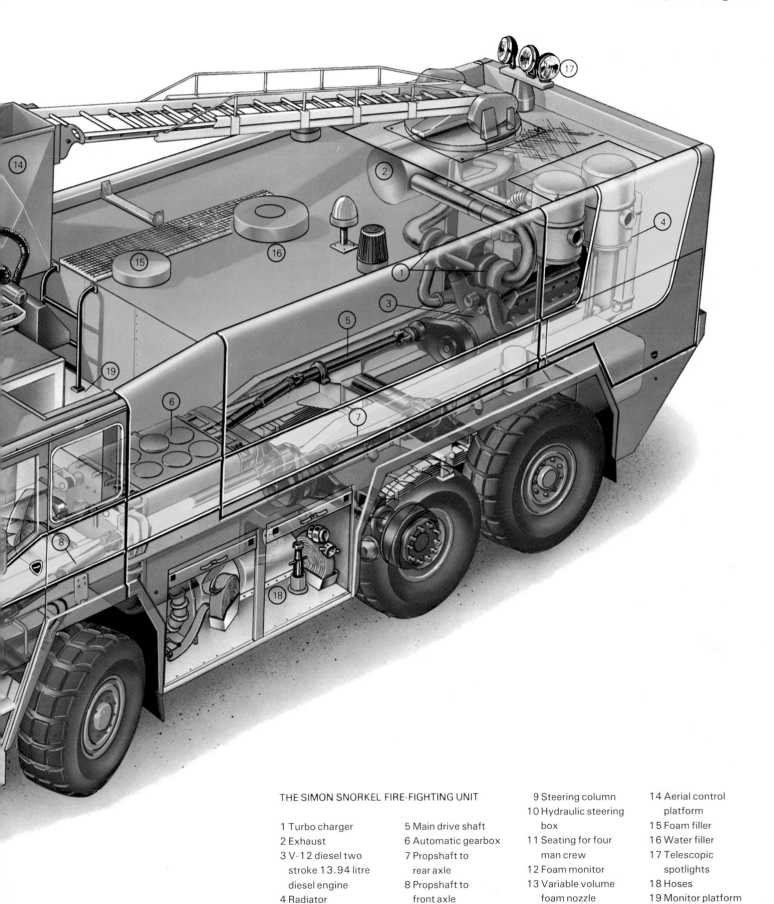

THE SIMON SNORKEL FIRE-FIGHTING UNIT

1 Turbo charger
2 Exhaust
3 V-12 diesel two stroke 13.94 litre diesel engine
4 Radiator

5 Main drive shaft
6 Automatic gearbox
7 Propshaft to rear axle
8 Propshaft to front axle

9 Steering column
10 Hydraulic steering box
11 Seating for four man crew
12 Foam monitor
13 Variable volume foam nozzle

14 Aerial control platform
15 Foam filler
16 Water filler
17 Telescopic spotlights
18 Hoses
19 Monitor platform

HELICOPTERS

Although the principles of helicopter flight have been known for centuries, the necessary technology and materials for the construction of a helicopter were not developed until the twentieth century.

The first known design for a helicopter-type aircraft was devised by Leonardo da Vinci, but a practical helicopter did not make a flight until 1936. This was the German Focke-Achgelis Fa 61. The first American helicopter was the VS-316A (later known as the R-4), designed by the Russian-born engineer Igor Sikorsky, which first flew in 1942.

The helicopter has two main advantages over the conventional, fixed-wing type of aircraft: it can land and take off vertically, rather than at a shallow angle; and it can hover motionless in the air, like a bird. In practice, this means that a helicopter can be used in many situations where conventional aircraft cannot go.

For example, a civilian helicopter can fly people right into the heart of a city centre, and land on the top of a building—no long runways, which have to be sited out of town, are needed. Military craft can fly men and machinery into the middle of the densest jungle.

The helicopter's hovering ability is probably most dramatically obvious when it is used as a rescue vehicle—as television series have shown. The helicopter can get right over the scene of the incident—whether it is on land or water, in open spaces or a crowded urban area—and stay above it, positioned with great accuracy. Rescuers and their equipment can be lowered by winch just where they are needed.

A conventional fixed wing aircraft is able to fly because of the lift generated on its wings as they move forward through the air. On a helicopter or autogyro, however, the fixed wings are replaced by a set of thin wings called blades attached to a shaft. The rotation of this set of blades, or *rotor*, through the air creates the lift necessary for flight.

An autogyro or a helicopter will climb when the total lift of the rotor exceeds the weight of the machine. The helicopter will hover when the sum of all the lift forces on the rotor blades is equal to the weight of the machine.

Although very similar, there is a technical difference between a helicopter and an autogyro. A helicopter has a rotor which is driven by an engine, but an autogyro has a rotor which gets its power from the motion of the airstream blowing through it, rather like a windmill. Thus the autogyro needs some other device, usually an engine-drive propeller, to pull or push it through the air horizontally.

Angle of attack

As the rotor turns, it traces out a circle in the air which is known as the *rotor disc*. To make the machine climb, the lift generated by each blade must be increased. This is done by increasing the angle of attack—the angle at which the leading edge meets the air-stream—of each blade equally, thus increasing the total lift without changing the direction in which it acts. The pilot controls this by means of a lever known as the collective pitch control.

In order to make the helicopter fly forwards the rotor disc

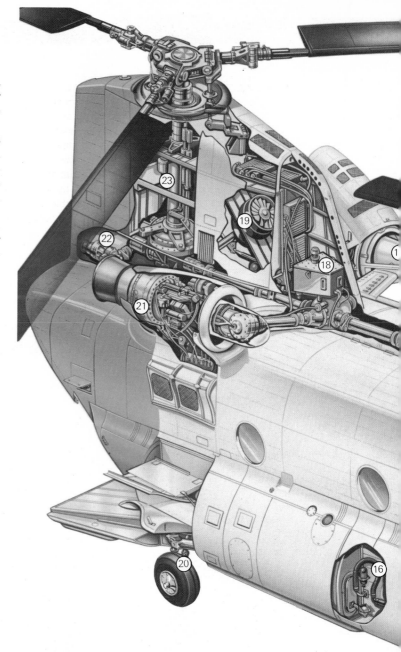

must be tilted forwards slightly, so that part of the rotor acts to pull the machine in that direction. The rotor disc is tilted forwards by increasing the angle of attack of each blade as it travels around the rear of the disc, and decreasing the angle as each blade travels around the front of the disc. As a change in the angle of attack means a change in lift, the lift is increased at the rear of the rotor disc and decreased at the front, causing the disc to tilt forwards very slightly.

These changes in the angle of attack of the blades can be made to occur at any point around the rotor disc, tilting the disc accordingly. This enables the helicopter to fly in any direction. The pilot controls the tilt of the rotor disc by means of the *cyclic pitch control lever*.

When a helicopter is moving forwards, the speed of a blade through the air changes as it travels around the rotor disc.

THE BOEING CHINOOK HC MK 1

1 Heated pilot tubes to measure speed
2 Cover for vibration absorbers
3 IFF aerial
4 Yaw sensing ports
5 Cyclic stick grip with speed trim and winch control switches
6 Rotor hub and oil tank
7 Air inlet to heater and blower
8 Jettisonable two-piece entrance door
9 Hydraulic rescue hoist (600lb strain)
10 Transformers, rectifiers and generators
11 VHF (AM)/UHF (AM) aerial
12 Troop seats (33 in all)
13 Fixed non-swivelling undercarriage
14 Trailing-edge trim tab
15 Forward drive synchronizing shaft
16 Fire extinguisher (10 each side)
17 Engine intake protective grill
18 Combined gearbox oil tank
19 Oil cooling fan
20 Fully steerable hydraulic undercarriage
21 Lycoming T55-L-11CS/SE engines
22 APU (power for engine starting)
23 Vertical drive shaft to rear rotor

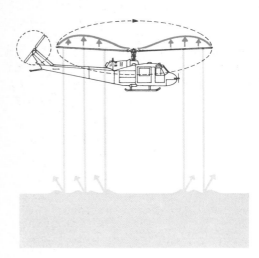

The helicopter produces the vertical force needed to lift it from the ground by forcing air downwards creating the down draught which is making the circular ripple patterns here. The down draught is strongest (arrows) beneath the points on the rotor at which most lift is generated.

When a blade is travelling towards the front of the disc, its air speed is its speed due to its rotation *plus* the speed of the aircraft.

But when the blade is travelling towards the rear of the disc, its air speed is the speed of rotation *minus* the speed of the aircraft. As the lift on a blade varies according to its speed through the air, the forward-moving blade has more lift than the rearward-moving one. These differences in lift would cause the rotor disc to tilt towards the rearward-moving side, and the helicopter would tend to fly sideways instead of straight ahead. To overcome this problem, the rotor blades are hinged at the roots to allow them to flap up and down a certain amount. This allows the forward-moving blade to rise slightly, in effect reducing its lift. The blades thus rise when moving forwards and drop again when moving backwards in order to keep the actual lift on them constant.

Torque compensation

The main, theoretical, problem with a helicopter is that it should not fly straight at all! According to the basic laws of mechanics, once off the ground the fuselage should spin round just as much (but in the opposite direction) as the rotor does. To prevent this a torque, or twisting force, compensating system is installed, such as the small anti-torque rotor at the tail of single main rotor helicopters.

Directional control can be achieved by varying the amount of torque compensation applied. Over-compensating turns the fuselage in the same direction as the main rotor, and under-compensating allows it to turn in the opposite direction. Where compensation is by means of an anti-torque rotor, the amount of compensation is controlled by varying the pitch, or angle of attack, of the blades.

Helicopters with twin main rotors, mounted one at the front and one at the back of the fuselage, do not suffer from the problem—they have a built-in compensating system.

Helicopter rotors are usually driven through a shaft fitted to the rotor hub (shaft drive), but some have been built with small jet thrust units fitted at the rotor blade tips (tip drive).

Shaft drive rotors can be driven by any form of aero engine. Originally all helicopters were powered by piston engines, but the turboshaft, or gas turbine, engine is now used on all but the smallest machines. Rotary engines, like the Wankel engine invented in Germany, are being considered for these small machines.

Autorotation

In the event of engine failure, the rotor rapidly slows down and loses lift, but it is possible for the pilot to land safely by use of autorotation. By rapidly lowering the collective pitch lever the pilot can set the blades so that their leading edges are pointing slightly downwards from the horizontal. As the aircraft is descending, the new position of the blades means that a positive, or upwards, angle of attack is maintained against the upward flow of the airstream. This generates forces on the blades to keep them spinning, and as the helicopter nears the ground the pilot raises the collective pitch lever slightly, so that the spinning rotor provides enough lift to slow down the machine before it lands.

The conventional helicopter cannot fly at more than about 400km/h (250mph) because at high speeds the air speed of the forward-moving blade approaches the speed of sound and that of the rearward-moving blade is very low. The result of this is that the rotor begins to lose lift: at a critical speed the blades will stall.

A partial solution to this problem has been put forward by the Sikorsky Company. This is the Advancing Blade Concept (ABC) which uses two identical rotors, positioned one above the other and turning in opposite directions.

HIGH-SPEED TRAINS

Look into any imaginary city of the future and, high above the spacious plazas, you will see the gleaming monorail cars gliding to and fro, whisking passengers swiftly and silently along the gracefully curving tracks. Present-day trains are slow and cumbersome by comparison. What chance is there of speeding them up?

In this high-speed world, time is a valuable commodity and no one wants to spend more time on a train than necessary. Rail experts recently began to realize that the majority of people value time so highly that they are prepared to pay for reduced journey times.

So researchers all over the world have been developing trains that will go faster. Japan's Shinkansen 'bullet' train, and the Italian Fiat and the Spanish high-speed train started to operate regularly in the late 1970s; Britain has its Advanced Passenger Train (APT), and there are equivalent French and Russian designs too. All these trains are aerodynamically styled to give maximum speed for the minimum energy consumption and have revolutionary suspension and bogie arrangements designed with the aid of computers. Rather than designing engine and coaches separately, the trains have been conceived as complete entities and probably more effort has gone into designing the chassis than the power unit. This is important: high-speed power units have been available for many years (way back in 1955 two French electric locomotives reached 331km/h, 205mph) but these speeds could not be maintained in regular service.

One of the main problems has been one of maintaining high speeds along twisty and steep sections of track without undue discomfort to passengers. Surprisingly, it *is* the comfort of passengers and not cornering ability that limits the speed of trains on curves—conventional trains could already run through curves much faster than they do, but only at the risk of flinging passengers violently from side to side. One solution is to build new tracks and eliminate the curves; this is what the French and Japanese have done for their new trains and there are considerable advantages in this solution. It separates the high-speed trains from slower traffic, for instance, and makes continuous operation at maximum speed easier.

However, building completely new lines is very expensive and often not possible. So Italian, Spanish and British researchers have developed trains that can travel at very high speeds on existing tracks. Their trains incorporate a hydraulic jack mechanism that tilts the body of the train automatically inwards on curves by up to nine degrees; the right amount of tilt for the curve and speed being set electronically. This helps to balance out the centrifugal forces generated in cornering and ensures the passengers a comfortable and stable ride. As a result, these trains can take bends up to 40 per cent faster than conventional trains—so the average speed is higher even though the top speed is not all that remarkable.

But conventional trains are not the only way forward, and there are many researches into less orthodox railways. In the United States, for instance, there is a suggestion that major cities such as New York and Washington are linked by a deep level 'atmospheric' railway. The idea is that air would be pumped out of the tunnel ahead of the train so that the train would be propelled by the pressure of air behind at speeds of up to 800km/h (500mph). The advantage of this system is that the pumps would use less energy than a train in moving air out of the way and so high speeds should be cheap. But perhaps the most exciting development is a train that will not run on wheels at all. Instead it will float a few millimetres above the rails, 'levitated' by power sources

One of the new breed of high-speed trains. This French TGV is a development of a conventional train, designed to run on new track laid without tight curves.

Right: One of the design features which helps the British Advanced Passenger Train (APT) to reach high speeds is its special tilt chassis, which leans the body of the train into bends independently of the bogie. But this makes the problem of picking up current from overhead wires even worse than it normally is. The problem was solved, and continuous high-speed pick-up improved by a special pantograph, which remains upright whatever the tilt of the body. An anti-tilt linkage isolates the pantograph from the bogie for all movements except roll and so keeps it directly above the centre of the track. Across the base of the mechanism there is a roll bar which is sufficiently flexible to absorb bumps, but rigid enough to absorb high cross winds.

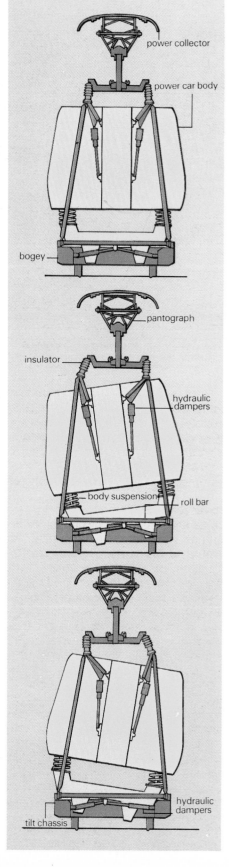

within either the track or the train itself. Here indeed is the gliding monorail of the writer's imagination.

Levitation trains are fast. Already prototypes have travelled at over 500km/h (320mph) and researchers are predicting even higher speeds (wheeled trains are unlikely ever to achieve much more than 350km/h, 215mph).

Achieving levitation

There are two basic methods of levitation—using a cushion of air (like a hovercraft) and using carefully regulated magnetic forces.

In the 1960s, hovercraft were fashionable and it was thought that air cushion vehicles had a bright future. Many countries invested in their own hovertrain projects: France had their Aerotrain; Britain had the RTV; Italy developed a couple of units at the Institute of Aeronautics in Palermo; there was the Transrapid 03 in Germany; and these were not the only examples. But by 1975, all these projects had collapsed and the way forward seemed to lie with magnetic levitation—*maglevs*.

Magnets can either attract or repel each other: some systems of maglev make use of the repulsion force; others use the attractive force.

The former system, known as *electrodynamic suspension* or simply EDS, is the type used in the German EET01 and the Japanese National Railway ML500. With EDS, there is a series of magnets in the underside of the vehicle and a series of coils in the track. With the vehicle moving the magnets induce an electric current in the track coils—the coils become electromagnets of the same polarity as the vehicle magnets and the repulsive force between the two magnets levitates the vehicle a little way above the track.

The train is driven along the track by another arrangement of coils and magnets, forming a *linear synchronous motor*. This works on the same principle as any electric motor and can indeed be thought of as a normal rotary motor rolled out flat. In the Japanese system, as the train moves along, it switches the power on and off in the section of track it is passing so that only a short stretch is electrified at any one time.

Try pushing two opposing magnets together smoothly in a straight line and you will appreciate that keeping a repelling magnet train perfectly on the track can present problems. Fortunately, a clever arrangement of the coils and magnets of the linear motor provides a simple and effective automatic guidance mechanism.

The mechanism is known as 'zero' or 'null flux' guidance. Because there is a vehicle magnet either side of the vertical track section, any deviation of the vehicle from a precisely central course means that one of the magnets must move closer to the track coils—this sets up a magnetic force which pulls the vehicle back in line. Only when the vehicle is on course is there no magnetic force and hence no sideways pull. There may also be a set of auxiliary rubber wheels either side of a central guide rail, to help with guiding at slow speeds.

The big advantage of the repelling magnet system of levitation is that there is no need for ride height regulation. Because the repelling force increases as the magnets of the levitation system move together, the vehicle stabilizes at a height in balance with its weight.

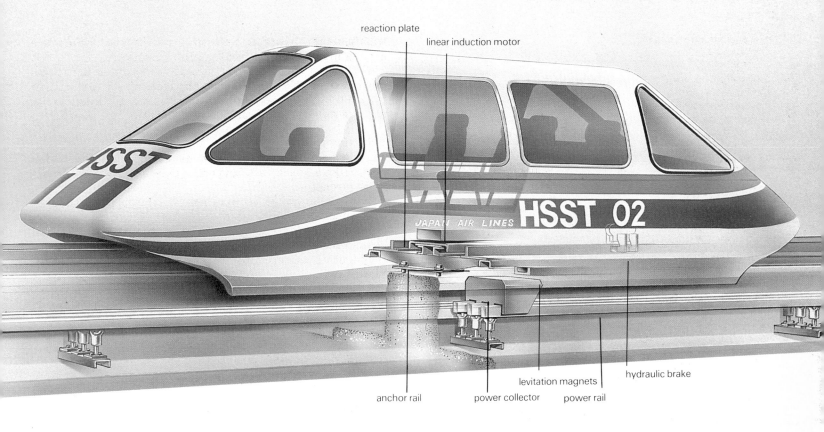

reaction plate

linear induction motor

HSST

HSST 02

JAPAN AIR LINES

anchor rail

power collector

levitation magnets

power rail

hydraulic brake

The big disadvantage of the system used in the ML500 is that the vehicle magnets have to be extremely powerful. And the only way to make them powerful enough is to use the science of *superconductivity*. The problem is that the magnets must be kept very cold for this effect to work. As a result, the ML500 project has had to develop special containers constructed like vacuum flasks and called cryostats to encircle the magnets with liquid helium at −269 degrees C. This adds very considerably to its weight and cost and to the maintenance problems of a maglev.

So, many people believe that the real future of maglev lies with the attractive system of *electromagnetic suspension* (EMS).

EMS has electromagnets in the vehicle and permanent magnets in the track. But the vehicle magnets are suspended from the chassis so that they lie below the track magnets. When the vehicle magnets are energized, they are attracted towards the track magnets and so pull the train clear of the ground. A special control system ensures that instead of making contact with each other, the magnets hang about 10mm (0.4in) apart, the attractive force being precisely balanced by the weight of the vehicle. Sensors continually measure the air gap and send information to a control system which regulates the current to the electromagnets, and gives precisely the right attractive force to maintain the air gap. In this way, the train hovers safely the correct distance above the track at all times. With such a small gap, and the fine control, the vehicle has no need of an undercarriage and comes to rest on a sprung glide mechanism.

Guidance is achieved in exactly the same way as levitation except that the guide rail and vehicle magnets are located vertically either side of the track.

As with EDS, EMS propulsion is provided by a linear induction motor. Because the electromagnets are in the vehicle, the vehicle has to be provided with electric current in any case, so it seems simpler to have the motor stator—that is, the powered part of the motor—in the vehicle as well. The stator, then, is only the length of the vehicle, and so this arrangement is called a 'short stator' motor.

Power problems

There are still major problems to be solved—supplying the vehicle with power, for example, and even designing the points in the track: points that have been designed are cumbersome and slow-acting and, above all, significantly increase journey times and reduce the traffic capacity of the network. In 1979, German researchers found they could not design a point to pass more than one train every five minutes. On conventional railways, the points can take a train every 1½ minutes. And it may be that the cost of a maglev system is just too much—and we will have to remain content with conventional railways.

On the other hand, speed is not the only advantage of a maglev. While Japanese National Railways have been aiming at a high-speed long distance maglev, the most exciting possibilities are for short haul networks within big cities. Positive guidance and linear motors mean that the track layout can be far more flexible and fit easily into confined areas. The temporarily-shelved British maglev, for instance, was capable of taking curves with a radius of just 8m (24ft) and climbing gradients of at least 5 per cent. This is a great advantage in crowded urban areas, particularly as the freedom from vibration and low weight makes elevating the track relatively simple.

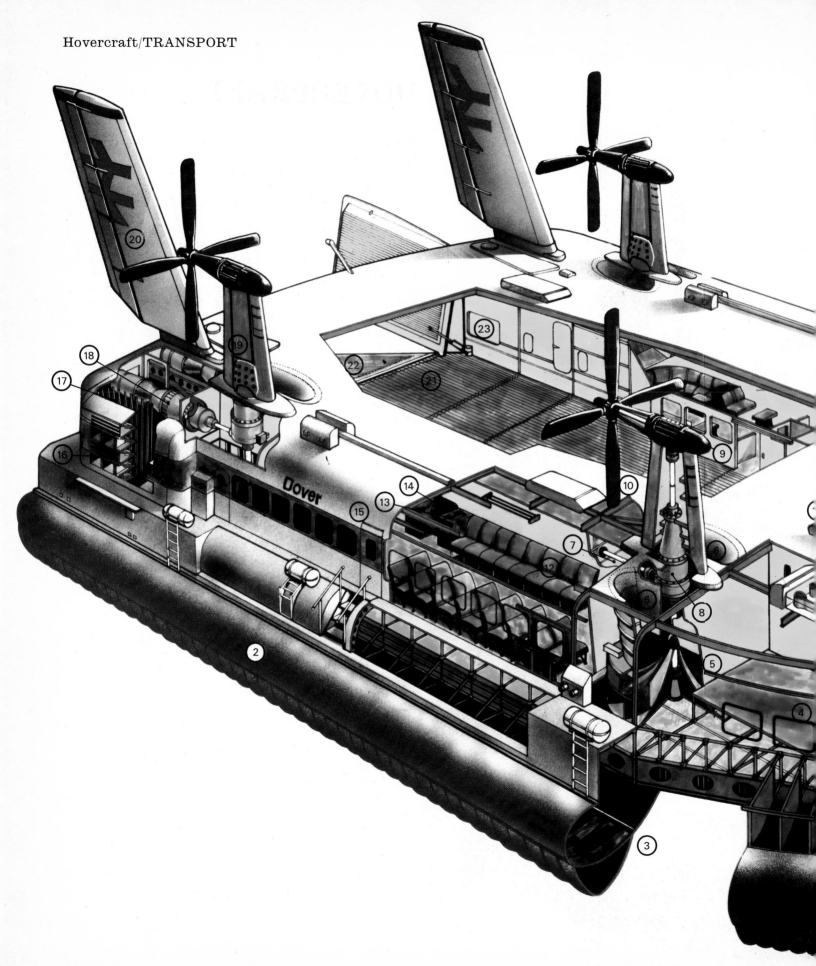

HOVERCRAFT

Until recently, no European had ever crossed the treacherous swamps that surround the upper reaches of the Amazon River in South America, yet a few years ago a British expedition not only made a crossing, but did so with a degree of ease and comfort barely conceivable to the explorers of old. Their epic journey was made possible by a remarkable but now familiar machine called the hovercraft.

A hovercraft has several advantages over ordinary land or water craft. First of all, it is both of these at once. It can cross the sea and run up the beach without stopping, as well as travel overland on any reasonably flat surface. Swamps are no problem, nor are waterways blocked with reeds, pack ice or other floating obstructions. A hovercraft can even go upstream against rapids, provided that they are not too steep, because it is unaffected by the speed of the current. It needs no landing strip or dock. It can just 'sit down' on any flat surface and the passengers can disembark on dry land or into boats as appropriate.

However, hovercraft are unlikely ever to take over completely from ships—there are various real drawbacks to the hovercraft options.

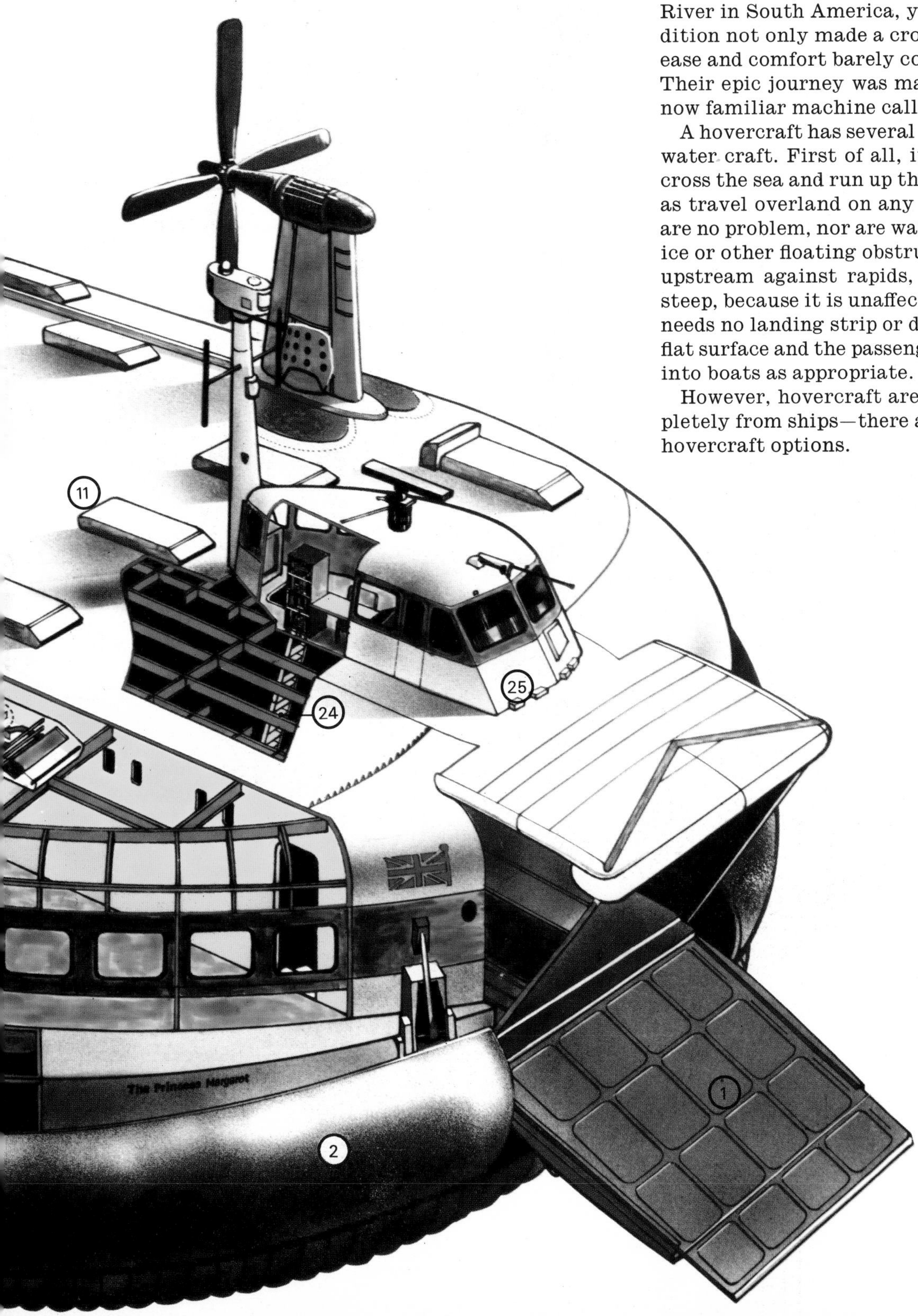

BRITISH HOVERCRAFT
CORPORATION SR-N4

1 Forward car ramp
2 Flexible skirt
3 Skirt fingers
4 Forward passenger compartment
5 Twelve-blade lift fan
6 Air intakes
7 Extension shaft from turbine
8 Main bevel gear box
9 Propeller gear box
10 HS Dynamics propeller
11 Air-conditioning packs
12 Main passenger compartment
13 Baggage racks
14 Door to car deck
15 Passenger entrance door
16 Engine intakes
17 Acoustic baffles
18 Marine Proteus gas turbine
19 Pylon
20 Fin
21 Car deck
22 Rear car ramp
23 Hatch to auxiliary power unit
24 Crew entry ladder
25 Control deck

An invaluable landing craft, the Bell Aerospace GEV can transport infantry and tanks to the heart of the action over surfaces, such as sand and swamp, that might halt other craft.

Although a hovercraft can travel fast using much less power than a ship of the same size, it always uses the same amount of power to lift itself no matter how slowly it goes. So at low speeds conventional ships and land vehicles are more economical. And even at fairly high speeds on water, a hydrofoil boat (see pages 171 to 173) uses less power—though a hydrofoil boat cannot be built as large as a hovercraft and cannot cross obstructed water. Neither can hovercraft cope with rough seas. The practical limit for a large hovercraft is a wave height of about 3m (10ft). Nor, on land, can they cross rocky ground or wide ditches, or climb even moderately steep slopes.

How a hovercraft moves

A hovercraft is supported above the ground by a 'cushion' of slightly compressed air which is blown under the craft by fans through a double ring of inward-angled air jets set all around the edge. In large craft the fans are driven by gas-turbine engines, in smaller ones by piston engines. The air cushion constantly escapes around the edge of the hull and has to be replenished. This loss is partly minimized by the angle of the jets. It is further reduced by a flexible rubber skirt which surrounds the edge of the hull.

This skirt traps air inside it and also acts as an extension of the bottom edge of the hull so that the solid bottom of the craft can rise, in the case of a large craft, 2m (6ft) or more clear of the ground. There is only a small gap under the skirt. This does not matter because the skirt bends to ride over waves and solid obstacles at the price of a very small increase in friction. Early hovercrafts were frequently out of action because of torn skirts but the current design has a skirt divided into short easily replaced lengths known as *fingers*.

Lifting a hovercraft is only one part of the problem—it also has to be driven forward. A ship-type propeller sticking down into the water would work perfectly well assuming that the craft never came on land. But it sacrifices the hovercraft's special advantage of being able to pass over floating obstructions and being equally at home on land and sea. So nearly all hovercraft use airscrews (like those of planes).

Small pleasure craft tend to have separate lift and propulsion engines. This makes it simple to control thrust (ie how hard the screws 'push', and so how fast the craft goes forward) by varying the speed of the propulsion engine independently of the lift engine.

On a big hovercraft, the airscrews are driven by the same engines as the lift fans. The engines turn at a constant speed to suit the lift fans, and drive both fans and propellers through fixed gearing. Thrust is varied by altering the pitch of the propeller blades: the variable-pitch blades can even be turned right round to give reverse thrust for travelling backwards and to slow the craft down. In a real emergency a hovercraft can stop very quickly by simply cutting the engines so that it does a 'belly-flop'. This is likely to tear the skirt, however, and would be damaging on land.

Steering is yet another problem—simply because there is so little friction between a hovercraft and the surface. Any attempt at a tight turn at speed results in the craft gliding off sideways. You could never drive even a small hovercraft along a road—it would drift constantly. Side winds blow it off course. The SNR4—a large craft from the British Hovercraft Corporation—has three steering devices which all work together: aircraft-type rudders at the tail, four propellers swivelling individually on pylons (so that it can move sideways if need be) and an arrangement for giving the propellers more thrust on one or the other side of the craft. In addition, the entire craft is made to tilt inwards on turns. However, it is still a handful to steer in windy weather.

Soviet designers are developing a completely different type of hovercraft which has no lift fan, but relies on lift created by its forward motion. It depends on the fact that an ordinary aircraft wing generates more lift than normal when it is very close to the ground or water surface. In effect it creates an air cushion under it, while also producing lift in the normal way by means of its curved top surface. The suggested design looks like an old-fashioned twin-engined, propeller-driven flying boat, but with very short, thick wings. At moderate speed—probably about as fast as the faster Western-type hovercraft—the hull lifts completely clear of the water and the craft skims over the surface. It will not actually take off because any greater altitude would lose it the extra lift created by the wings' closeness to the surface. It would be free of the disadvantages of the easily damaged skirt of an ordinary hovercraft, though it would suffer from the same steering problems.

HYDROFOIL

For most of the last 150 years, shipowners have watched helplessly as the speeds of new and competitive forms of transport increased tenfold, while that of their own vessels barely doubled. The first passenger-carrying aircraft were lucky to make 100km/h (60mph). Today most commercial jetliners operate ten times faster, and Concorde is 20 times faster. By contrast, the ocean-going liner Queen Elizabeth II averages 30 knots—little more than twice as fast as the 14½ knots achieved by the Great Eastern in 1858. What can be done to speed up ships?

The problem is that a displacement vessel, with most of its hull below water, must force its way through the sea against wave motion which resists its progress. Even more difficult to overcome, a friction known as 'skin effect' is created between hull and sea which increases with the speed and displacement of the vessel.

The solution, simple to express but difficult to achieve, is to lift the ship sufficiently for its hull to skim over, not through, the waves thus allowing energy previously spent on overcoming wave resistance and skin effect to be used directly for developing speed. The pursuit of this goal has led to two very different types of craft, known as *surface skimmers*—the spectacular hovercraft (see page 168) and the less dramatic, but more practical, hydrofoil.

A hydrofoil achieves its lift by means of specially designed 'wings' or foils attached to the hull of the boat—see diagrams overleaf. As the craft gathers speed it begins to rise in the water until, when it is travelling fast, the lift created by the flow of water over the foils is sufficient to raise the hull clear of the water. A hydrofoil boat in full flight looks rather like a seaplane taking off, and behaves similarly.

The lift is created by the application of the *Bernoulli principle*. The eighteenth century Swiss scientist, Dr. Daniel Bernoulli, in his studies of gases and fluids in motion, showed that the faster a fluid or gas moves, the lower the pressure it exerts upon objects along which it flows.

The shape of the foil causes water to flow at a higher speed over its curved upper surface than over its flat, or even slightly concave, lower surface. The water passing over the upper surface has to travel further, and therefore faster, than that passing beneath. This causes a reduction in pressure on the upper surface, thus creating the 'lift' which raises and then maintains the craft above the surface of the water. Aircraft take off and fly by exactly the same effect.

Foil types

There are two main types of foil: the *fully-submerged* system (which permits foil-borne operations in almost any ocean conditions); and the *vee-foil surface-piercing system*. The two concepts both aim at overcoming one basic problem: that the upper and lower depth limits within which foils can operate successfully are too fine to permit manual control.

In fully submerged hydrofoils, the foils are completely immersed in water all the time. The amount of lift produced is controlled in modern craft by using hydraulic rams to alter the angle of attack of the foils. The rams themselves are directed by signals from a sonic device in the bows, which sends out pulses of high-frequency sound waves in order to assess their height and thus the correct angle of foil inclination. Gyroscopes in the hull sense the craft's pitch, roll and heave motions, and feed this information into the same control system to keep the boat level. These very expensive systems are used mainly in naval hydrofoils.

Fully submerged foils are located at depths where the action of the waves does not greatly affect them. But, since they have no natural stability, an additional control system is required to prevent the top-heavy craft from crashing down on to its hull. Some advance warning of the sea state immediately ahead is needed if a constant hull attitude is to be maintained.

In the surface-piercing system, part of the foils actually breaks the surface of the water as the craft moves forward through the peaks of waves. A drop in lift occurs because the pressure difference between the upper and lower surfaces is destroyed when the water no longer flows over the top. The

Right: A Soviet-built hydrofoil leaves Piraeus, carrying holidaymakers to the Greek Islands, and outpacing traditional ferries. Like a hovercraft, a hydrofoil glides above, rather than through, the water—but in fact a hydrofoil operates much more like an aircraft than does a hovercraft. It has been described as a cross between the ship it looks like and the aircraft it is built like.

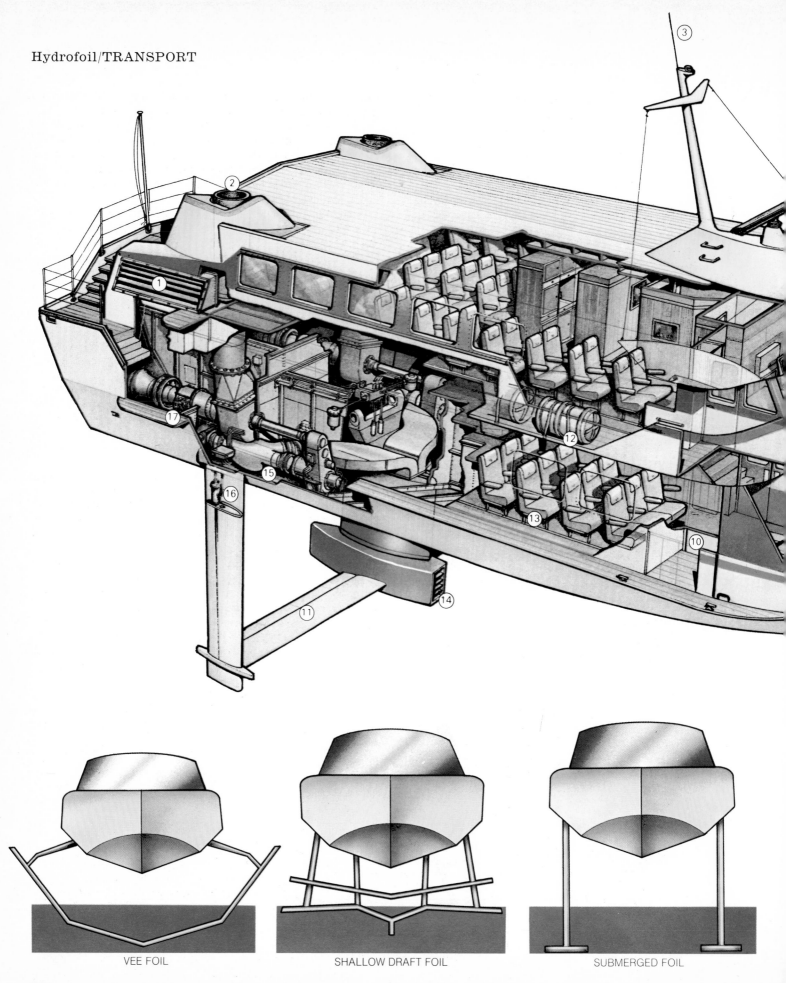

VEE FOIL

SHALLOW DRAFT FOIL

SUBMERGED FOIL

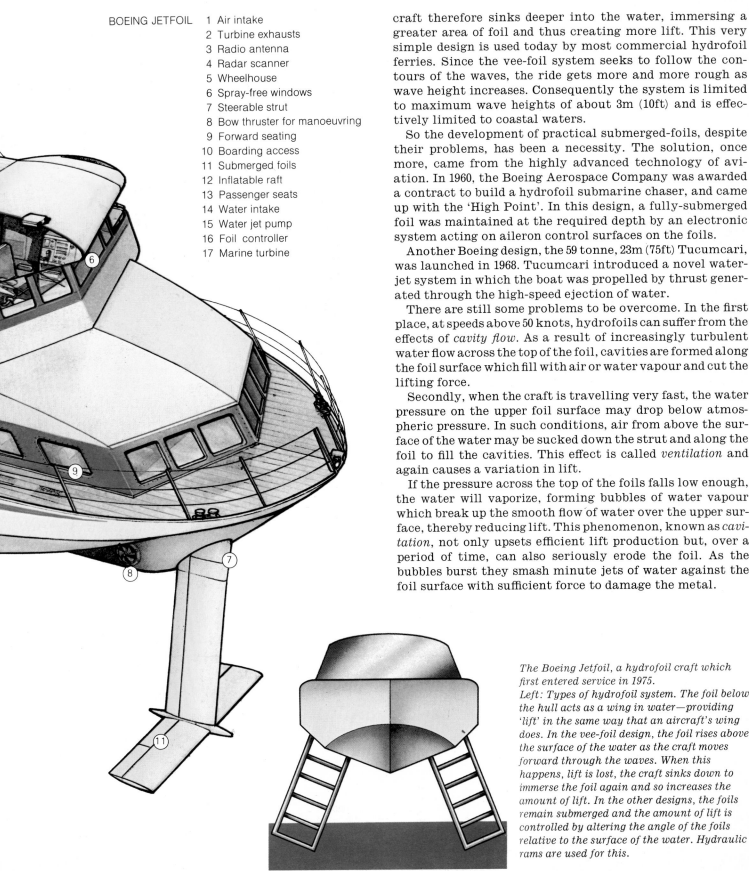

BOEING JETFOIL
1 Air intake
2 Turbine exhausts
3 Radio antenna
4 Radar scanner
5 Wheelhouse
6 Spray-free windows
7 Steerable strut
8 Bow thruster for manoeuvring
9 Forward seating
10 Boarding access
11 Submerged foils
12 Inflatable raft
13 Passenger seats
14 Water intake
15 Water jet pump
16 Foil controller
17 Marine turbine

LADDER FOIL

craft therefore sinks deeper into the water, immersing a greater area of foil and thus creating more lift. This very simple design is used today by most commercial hydrofoil ferries. Since the vee-foil system seeks to follow the contours of the waves, the ride gets more and more rough as wave height increases. Consequently the system is limited to maximum wave heights of about 3m (10ft) and is effectively limited to coastal waters.

So the development of practical submerged-foils, despite their problems, has been a necessity. The solution, once more, came from the highly advanced technology of aviation. In 1960, the Boeing Aerospace Company was awarded a contract to build a hydrofoil submarine chaser, and came up with the 'High Point'. In this design, a fully-submerged foil was maintained at the required depth by an electronic system acting on aileron control surfaces on the foils.

Another Boeing design, the 59 tonne, 23m (75ft) Tucumcari, was launched in 1968. Tucumcari introduced a novel water-jet system in which the boat was propelled by thrust generated through the high-speed ejection of water.

There are still some problems to be overcome. In the first place, at speeds above 50 knots, hydrofoils can suffer from the effects of *cavity flow*. As a result of increasingly turbulent water flow across the top of the foil, cavities are formed along the foil surface which fill with air or water vapour and cut the lifting force.

Secondly, when the craft is travelling very fast, the water pressure on the upper foil surface may drop below atmospheric pressure. In such conditions, air from above the surface of the water may be sucked down the strut and along the foil to fill the cavities. This effect is called *ventilation* and again causes a variation in lift.

If the pressure across the top of the foils falls low enough, the water will vaporize, forming bubbles of water vapour which break up the smooth flow of water over the upper surface, thereby reducing lift. This phenomenon, known as *cavitation*, not only upsets efficient lift production but, over a period of time, can also seriously erode the foil. As the bubbles burst they smash minute jets of water against the foil surface with sufficient force to damage the metal.

The Boeing Jetfoil, a hydrofoil craft which first entered service in 1975.
Left: Types of hydrofoil system. The foil below the hull acts as a wing in water—providing 'lift' in the same way that an aircraft's wing does. In the vee-foil design, the foil rises above the surface of the water as the craft moves forward through the waves. When this happens, lift is lost, the craft sinks down to immerse the foil again and so increases the amount of lift. In the other designs, the foils remain submerged and the amount of lift is controlled by altering the angle of the foils relative to the surface of the water. Hydraulic rams are used for this.

HYDROGEN CAR

It has been called 'the forgotten fuel'. Enthusiasts claim that it could end the energy crisis at a stroke, powering aircraft and road transport for the foreseeable future. Hydrogen has been 'the fuel of the future' ever since a Jules Verne character first extolled its virtues. It burns at a high temperature, contains much more energy than a similar weight of petrol, and it creates far less atmosphere pollution.

Unlike petrol, there is no free (uncombined) hydrogen in the Earth, and only a little in the atmosphere. But there is no shortage of the raw materials required to produce it. All that is needed is water, plus energy in some other form—and there are many promising processes being developed.

The design of the petrol-powered internal combustion engine requires relatively few changes before it is suitable for use with hydrogen as its fuel. Hydrogen burns in air about twelve times faster than the present hydrocarbon-air mixture. In some experimental engines, the flame flashed back to the carburettor (where hydrogen and air are mixed in the usual way) when the engine was loaded by rapid acceleration or hill-climbing. Fortunately, microprocessors can be used to control the timing of the ignition spark more accurately.

One advantage of a hydrogen engine is that pollution is reduced. There are no unburned hydrocarbons emitted, but in the heat of the hydrogen reaction some nitrogen from the air may still combine with oxygen to form oxides of nitrogen. One solution has been to inject water, which vaporizes in the cylinder as the hydrogen burns, and reduces the temperature to a level where the nitrogen reactions stop. At the same time, the water vapour adds bulk to the expanding gas in the piston: it actually helps the engine to deliver its power efficiently. In early vehicles of this type, a water tank had to be added and frequently topped up; in recent designs, water is recovered from the exhaust. With only one potential pollutant, hydrogen is far better than petrol, which produces several that are expensive to get rid of.

In an alternative approach to these problems, the engine has no carburettor. The same volume of air is drawn in for every stroke of the piston, and a metered amount of hydrogen is injected, the actual amount being controlled by the accelerator pedal.

The fuel tank is a greater problem. How can a car carry enough hydrogen gas to travel the 200 miles or more that we expect from a tank full of petrol? Carrying hydrogen in the form of gas is impractical. You would need to carry around dozens of large, heavy gas cylinders to achieve even a short range—more weight and bulk than even a small truck could carry.

Keeping it cool

So, is liquid hydrogen better? This has an impressive record of success in space—it was used to propel men to the Moon—but could it propel a car over our required two hundred miles? The answer is yes, but liquified hydrogen has to be kept at below minus 253 degrees C, which calls for sophisticated cooling mechanisms and vacuum flasks. And, for the same range, you need a bigger 'tank' for hydrogen than for petrol.

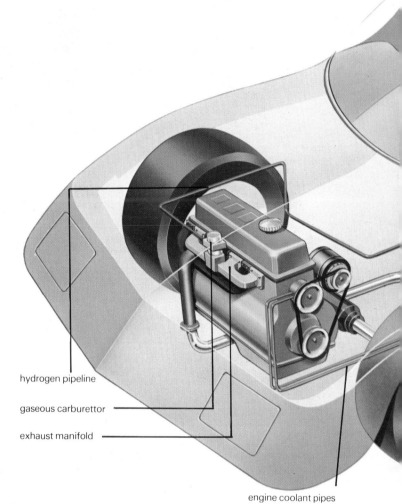

hydrogen pipeline

gaseous carburettor

exhaust manifold

engine coolant pipes

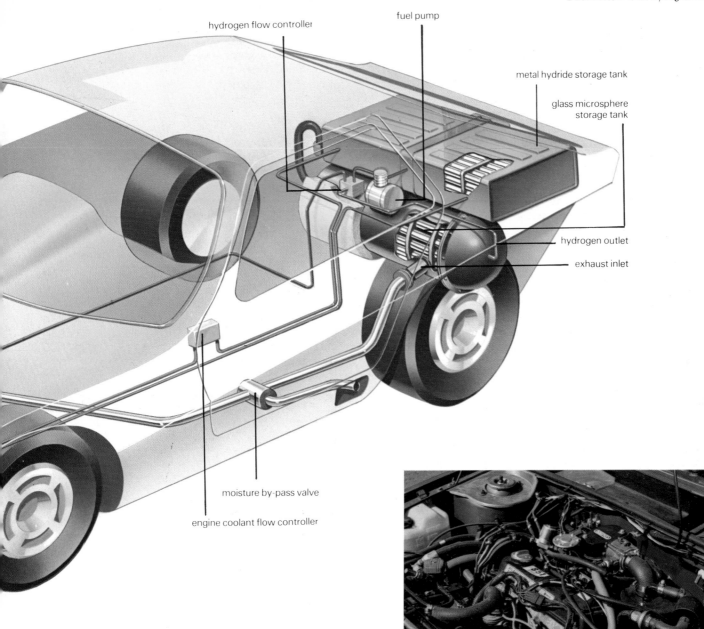

hydrogen flow controller

fuel pump

metal hydride storage tank

glass microsphere storage tank

hydrogen outlet

exhaust inlet

moisture by-pass valve

engine coolant flow controller

So most experts have favoured another approach—the *metal hydride*. Grains of certain metallic alloys, notably of iron and titanium can absorb vast amounts of hydrogen at room temperature, giving off heat as they do so. The process is reversible—heat the metal hydride and hydrogen emits.

Although the metal (in its containing tank) is still heavy, it is much lighter than gas cylinders. An acceptable weight would give a vehicle a range of about 115km (70 miles). Recharging a hydride cell is at best a much slower business than filling up a petrol tank. And with long use, the grains of metal may gradually break up, forming a fine powder which obstructs the flow of gas. There is plenty of room for improvement.

However, the breakthrough may come in an entirely different technology—using a *fuel cell*, in which hydrogen combines with oxygen to produce electricity very efficiently. Instead of an internal combustion engine, such a hydrogen car would use an electric motor.

Opposite: The hydrogen-powered car may one day be a practical proposition. Daimler-Benz are already testing prototypes.

Above: This may look like a conventional car engine, but in fact it has been adapted to run on hydrogen fuel. Only minor changes to carburation and the cylinder head are needed to a standard internal combustion engine.

IN-CAR COMPUTERS

Every minute of the day someone, somewhere, dies in a road accident because a driver has made a human but fatal mistake. With more traffic on the roads today than ever, many traffic engineers long for the day when each car will be controlled by a computer and the driver can sit back in his seat knowing that the computer will steer him safely clear of all potential accidents—and traffic jams. What are the possibilities for in-car computers?

Ideally, of course, cars would be totally automatic. Information supplied by cables buried in the road would be received by the car's central computer terminal, and the car would then be automatically guided to its destination swiftly and safely. All the driver would have to do is switch on the car and tell it where he wanted to go.

Much of the basic technology for this kind of control already exists. But, for the foreseeable future at least, silicon chip microprocesses will be used to monitor and control just a few specific functions, on a conventionally driven car.

Providing the information

With the advent of the chip and its continuous refinement, there is no major technical barrier to equipping a car with a central computer small enough to be practical but sufficiently powerful to allow complete automation of the car's functions. However, the large number of functions in the average car and the larger number of different factors which affect their operation make providing the information for a computer a less than simple task.

The microprocessor for a fuel injection system, for instance, needs to have information on engine speed and load, throttle setting, air temperature and air density, if it is to provide the correct mixture under all conditions—and even this may not be enough. Each of these quantities requires a separate sensor, each of which must be tied in ('interfaced') with the computer. Whether the sensor registers as a varying electrical resistance, as a signal from a photocell or whatever, it must be translated into a form that the computer can understand.

Unfortunately, information from sensors is rarely in the digital language that computers operate in—it is usually in 'analogue' form. Converting from analogue to digital form adds further to the complexity and expense.

There is the outward interface, too—the one that changes the computer's decisions into actions on the car. This usually means first amplifying the output of the chip by a large degree. Then there needs to be some way of changing the electrical digital signal into a mechanical analogue one so that the signal can, for instance, operate the car's brakes. This is not usually too difficult to achieve.

A greater difficulty is avoiding the setting up of unwanted oscillations. This is because the control system forms a closed-loop—the brakes (or whatever) send a signal to the computer; the computer sends a signal back to the brakes, telling them to ease off or brake harder. So the signal that the brakes are sending alters, and this in turn alters the computer's output signal. If the system is designed and adjusted correctly, then the difference between what the brakes are doing, and what the computer wants them (the 'error') to do will get progressively smaller as the information and command signals messages continue. But in some cases, the computer might over-correct the error—the result is that the error will get larger instead of smaller, and the system will go into oscillation—probably fatal if it occurs in a braking system.

It is clear that the problems of interfacing are far more of an obstacle to the development of the computerized car than the computer itself. Computer systems are already well developed and their capabilities known and it is the perplexities of the interface that are absorbing the majority of the research into electronic systems for cars.

Electronic ignition

Development of electronic fuel supply and ignition controls has been given an extra boost by the introduction of strict laws (notably in the United States and Japan) governing the amount of polluting gases that may be present in the exhaust. The limits can be met only by controlling ignition timing and fuel delivery far more accurately than had previously been the case.

Many cars now have electronic ignition. Although most are still timed mechanically, a number of more recent systems have begun to incorporate microprocessors to improve the accuracy with which the ignition is timed to fire the spark plug. Data is supplied to the microprocessor concerning engine speed, the position of the piston (taken from the crankshaft position), and the load on the engine. The microprocessor then compares the information it receives with instructions imprinted in its memory during manufacture

How a computer might control a car. Sensors (3) are located throughout the car to monitor such things as oil pressure, engine temperature and the amount of fuel. The information they gather is fed to a 'subscriber station' (1) which turns the electrical analogue signals into digital form and transmits them to a central controller (2). This processes the information, and sends messages back to the subscriber stations which tell the various parts of the engine what to do.

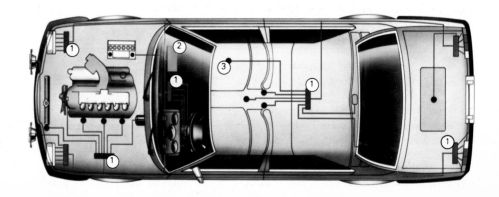

and the ignition is duly fired at the correct moment. This sort of facility had already been shown to improve performance and reduce toxic emissions.

These systems, however, are only a starting point and the potential for processor control is much greater. For instance, by the end of the 1970s a number of manufacturers had introduced a 'knock sensor'. This is valuable because engines could be more efficient if they ran on higher compression ratios. At present petrol engines rarely have compression ratios of more than 10 to 1, and restrictions on the lead content in petrol often make them much lower. If they were any higher, the engine would be subject to detonation or 'knocking'. When this occurs some of the mixture explodes violently before it is ignited rather than simply burning. Knocking not only wastes power, but can wreck the engine. The knock sensor detects knocking as it is about to start, since incipient knocking shows up as an unusual variation in crankshaft's speed before audible knocking begins. If knocking is detected, the processor retards the ignition to prevent it.

Fuel injection systems have been using microprocessors for some years. Originally varying the mixture only according to engine speed and throttle setting, many systems now include sensors for monitoring such things as engine temperature and air pressure. More recently, the German manufacturer Bosch introduced a 'lambda sensor' to detect the amount of oxygen in the exhaust (which is an indication of the efficiency of combustion). It works by checking the electrical conductivity of the exhaust gases. Any deviation from

One idea for a solid state instrument panel. Although the version shown is only an engineer's dream, the instrument panel is where you will find the most obvious signs of the presence of electronics in even today's cars. There is little advantage in electronic displays themselves—the conventional analogue dials show speed, engine rpm, fuel level, and so on, just as well—but they do open the way to the introduction of a vast array of other functions. Some cars already have a display which gives a continuous display of fuel consumption and average speed. And 'trip computers' can work out such things as the estimated time of arrival at your destination. Researchers suggest that a digital display can be projected onto the windscreen to give the driver all the information he needs without having to move his eyes from the road. These head-up displays (one giving speed, and a representation of stopping distance is shown in the picture above) have been familiar to the pilots of jet fighters for years. But so far, no one has been able to develop a format small enough or cheap enough for use in cars, or worked out how to provide the illumination needed so that displays will be bright enough to be always visible even in direct sunlight.

the set limits can then be corrected by altering the fuel injection settings via the microprocessor.

Electronic operation of the engine valves opens up a further, enormous, range of possibilities. It would, for instance, allow the driver to change the characteristics of his engine at will. At the flick of a switch, he could change the car from a highly 'tuned' sportster to an economical slogger with plenty of pulling power simply by altering the timing and duration of the valve opening. Or he could shut off the valves altogether, turning the engine into an air compressor for efficient engine braking.

The engine is by no means the only part of the car which can benefit from electronics. Another promising area is the transmission and some engineers believe that, in the near future at least, there is more to be gained in economy from controlling the transmission electronically than by applying microprocessors to the engine. In such a system, the microprocessor would have an imprint of the engine characteristics and would change the gear ratio automatically to match these with the prevailing speed and load conditions.

Anti-lock braking system

One area that has received only scant attention is the car's braking and suspension system. Now, however, BMW and Mercedes offer an anti-lock braking system on their top-of-the-range models. Considering that aircraft have had such systems for more than 20 years it is, perhaps, surprising that this development has taken such a long time to arrive. The trouble was that while a hydromechanical sensor and control system worked well enough for a big, heavy aircraft wheel, it did not respond quickly enough with the much lighter braking system of the average car. Electronics provided the answer. In an electronic anti-lock system, the sensor detects that the wheel is slowing down too quickly—that is, it is about to lock—and the circuitry and output releases the brakes quickly so that the wheel does not lock and then equally quickly re-applies them.

Some engineers feel that if the brakes can be controlled in this way through their hydraulic pipes, then so could a car's suspension system. Systems have already been demonstrated which prevent a car rolling on bends, but a fully electronic system could do much more. Not only roll, but also 'pitch'—end-to-end rocking of the car due to braking and acceleration—could be controlled, and the quality of the ride could be varied to suit the driver. Most conventional car suspension systems are restricted by the design of their spring. An ideal spring should act progressively, being soft at the beginning of its travel to absorb small bumps, and becoming steadily stiffer towards the end to react to large bumps. This is difficult to achieve with mechanical springs, but less so with high pressure air/liquid springs and these, in conjunction with a microprocessor controlling a valve to either let air out of the spring or pump it in, could give excellent suspension.

With microprocessors playing an increasing role in the control of individual functions, now is the time to assess the chances of all these separate systems becoming linked together and governed by one central computer in a fully automated car.

Although a full scale electronic car has not yet been built, there are signs that the concept is not as far off as might be thought. Once all the car's functions come under computer

control, it becomes more feasible to allow the computers to react to external features such as traffic and route deviations. One possibility is the incorporation of a microwave radar linked to the computer. This would detect the positions and speeds of other cars and pedestrians, even in dense fog. It could then link automatically with the car's braking system to keep a safe distance or stop. It could also tell the driver whether it was safe to overtake on twisty roads by 'seeing' round the bend.

Already by the late 1970s, a number of people, particularly in West Germany, were trying out systems whereby information about traffic conditions and alternative routes from a central computer could be picked up from the road and displayed on a console within the car. And it needs only one step from displaying the information to getting the car to react directly—fully automatic guidance systems are perhaps not all that far away.

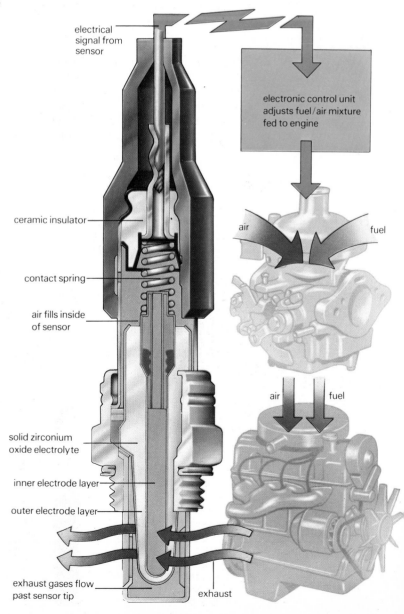

electrical signal from sensor

electronic control unit adjusts fuel/air mixture fed to engine

air   fuel

ceramic insulator

air   fuel

contact spring

air fills inside of sensor

solid zirconium oxide electrolyte

inner electrode layer

outer electrode layer

exhaust gases flow past sensor tip

exhaust

An electronic fuel control system. At the heart is a sensor which produces an electrical signal in response to the amount of oxygen in the exhaust gas. The electronic control unit then alters the fuel/air mixture to give better economy and a cleaner exhaust.

SUPERBIKES

Stand beside a modern 'superbike' and count from one to 12. Such is the devastating acceleration of these machines that, by the time you have finished counting, it can be over a quarter of a mile away and travelling at more than 120mph. Performance like this has been attained by years of research and experiment.

Because the owner of the big bike wants from his machine power and performance more than anything, many of the manufacturers have devoted their immense resources and research facilities to developing smooth powerful engines—some riders say to the exclusion of everything else.

In the development of these engines the keynote has always been the search for power. It is this which has led to the big 4-stroke multi-cylinder, multi-valve, overhead cam engine as the norm for the typical superbike able to travel at more than 220km/h (137mph) without excessive vibration.

Power boost

More than any other single factor, this power boost has been obtained by allowing the engine to run faster. Power is the product of torque—the turning force exerted on the crank-shaft by the pistons pushing down—and the engine speed. So the faster the engine can run, the more power it can produce, provided the extra rpm are gained without losing much torque.

All these extra rpm have been squeezed out of what is still a basic 4-stroke engine by a combination of improvements in valve gear and the relationship between piston weight and its speed relative to engine revs.

The weight and speed of the piston are important because the piston, conrod and crankshaft assembly can take only so much stress. The amount of stress involved depends on the violence of the piston deceleration at the end of each stroke—that is, how fast it was moving before it was brought to an abrupt halt—and the weight of the assembly behind it, trying to keep it moving in the same direction. If either piston speed or the weight of the assembly can be appreciably reduced, the engine can run much faster. This is what the Japanese have done.

Unlike European manufacturers who have tended to stick with the more traditional layouts, they have concentrated on the multi-cylinder engine. Although these are normal on cars, they are a relatively recent innovation for motor cycles. The objection has always been that they are too heavy and too big for a bike, but they have a number of advantages when it comes to obtaining more power by getting

Above: The Kawasaki Z1300, one of the classic Japanese superbikes, and a good example of their philosophy for getting speed on two wheels. It has an advanced, 6-cylinder engine, which is water cooled.

Right: In contrast to the Kawasaki, this prototype of a British superbike, the 1000cc Hesketh, shows the direction that British designers have taken. The engine for this bike has only two cylinders, arranged in a V formation developed from the British CCM 500cc single cylinder engine design.

the engine to run faster. First of all, by dividing the engine into four or six cylinders, each of the pistons can be smaller and lighter for a given engine capacity. Secondly, they are more compatible with a short stroke (the stroke is the distance the piston travels in the cylinder from top to bottom) and a short stroke means lower piston speed since the piston has a shorter distance to travel for a given engine speed. Thirdly, the multi-cylinder engine runs very smoothly because each piston can be used to balance out some of the vibration caused by the others.

Valve gear

None of these improvements would have made much difference if they had not gone hand in hand with a corresponding development of the valve gear (the valves let fuel into the cylinder and exhaust gases out).

Until recently by far the majority of 4-stroke engines were 'ohv' (overhead valve) types, and many still are. This means that the valves are set on top of the engine (as they are in all modern engines) but the camshaft which controls their operation is not, and the two mechanisms must be connected by a series of pushrods and rockers. While easy to adjust and reliable, the pushrod mechanism adds considerable weight to the moving parts of the engine and, more importantly, begins to work badly at high engine speeds so that valve opening and closing is inaccurate. Although a few powerful new engines still use ohv, most are now equipped with the overhead camshaft (ohc) arrangement in which the camshaft is also located at the top of the engine. Not only are the pushrods dispensed with, but in many cases the rockers are as well. In the most sophisticated units there are two camshafts, one for the inlet valves and one for the exhaust. The

Italian Ducati's remarkable *desmodromic* system not only includes cams for opening the valves but also for closing them positively via a forked arm rather than simply returning them under spring pressure.

Overhead camshafts can operate faster, more quietly and accurately than ohv—though at the expense of ease and adjustment.

Ignition advances

Hand in hand with the increasing development of the motor-cycle engine as a sophisticated high output power plant have been equally impressive advances in carburation and ignition. The crude 'fixed jet' carburettors favoured by automobiles were abandoned in the motorcycle world many years ago, but, although the throttle slide units which replaced them gave considerably more accurate and even control of the fuel/air mixture, they were by no means perfect. Their failings were particularly obvious when the rider snapped open the throttle sharply and a marked weakening of the mixture gave a disconcerting power drop-off. By 1980, however, most of the big superbikes were equipped with one of the ingenious constant vacuum carburettors for each cylinder. These give constant and accurate metering of the mixture however clumsily the rider handles the throttle control. Nevertheless, even CV carburettors have their failings, and it may be that the future lies with fuel injection.

Again, in the ignition circuit the superbike shows its technological lead over the average car. While many cars still have the rather primitive coil and contact breaker system, superbikes have had electronic ignition as standard for a number of years and each year the systems are becoming more and more refined.

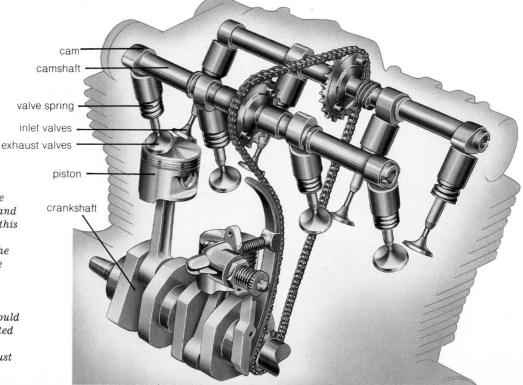

An advanced overhead camshaft gear. The valves of an engine move slightly to open and close holes in the cylinders (not shown in this drawing) in which the pistons move. The valves are pushed down by the action of the cams, and return by means of springs. The camshafts are mounted directly over the valves—with a conventional engine, the shafts would be mounted down by the crankshaft, and the motion of the cams would be transmitted to the valves by a complicated system of rods and levers. This engine has separate camshafts for the inlet and exhaust valves—a double overhead camshaft arrangement.

cam
camshaft
valve spring
inlet valves
exhaust valves
piston
crankshaft

To provide a spark at the spark plugs requires a high voltage impulse, passed at exactly the right moment. In the old contact-breaker system, control of the spark is carried out mechanically—but contact breaker points wear rapidly, and need frequent adjustment. In electronically triggered ignition systems, there are no mechanical parts to wear out. Instead, the control of the spark is carried out by an electronic switching unit. In turn, the switching unit is controlled by an optical trigger—a photocell sends an electrical signal to the switching unit as long as it is activated by a beam of light: the beam of light is covered and uncovered by a simple rotating shutter, called a chopper. In more sophisticated systems, the switching unit is replaced by an electronic voltage generator, which sends high voltage impulses direct to the spark plugs.

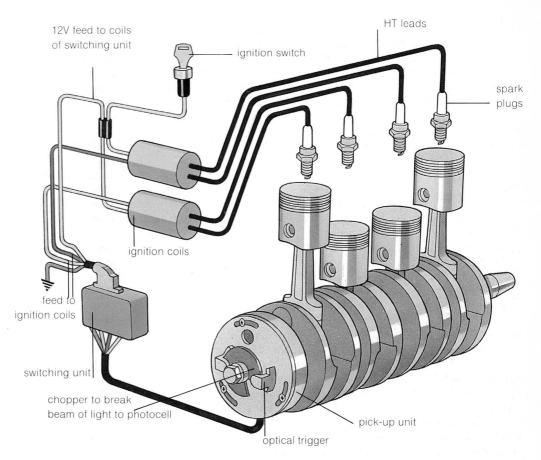

12V feed to coils of switching unit

ignition switch

HT leads

spark plugs

ignition coils

feed to ignition coils

switching unit

chopper to break beam of light to photocell

optical trigger

pick-up unit

A sophisticated engine is only the beginning—to produce a fast bike, this has to be allied to an equally sophisticated bike frame and suspension system. But modern bikes are often not very advanced in this area.

One area in particular that received less attention from the Japanese, until recently, is a bike's stability at high speeds and on corners. The factors involved are hard to pin down, and research is only just beginning to identify the reason that some bikes are superior to others in this respect. Nevertheless, it has been known for years that any flexibility in the frame and suspension is undesirable.

Rigidity in the frame is really achieved only by proper triangulation of straight frame tubes to distribute stress evenly. The conventional *duplex loop* frame with its large central box and many curved tubes is sadly lacking in this respect and good triangulation is rare in modern production bikes; the Italian Ducati also obtains extra strength by using the engine as a frame member—an old idea but one which may become more popular in the future and may be useful where the engine is wide and massive, like the Japanese fours, and difficult to set in a triangulated frame.

Experiments at Manchester University in England have highlighted a number of other causes of high speed instability and poor handling. Poor weight distribution and suspension weaknesses are especially significant. A heavy bike, or one with a high centre of gravity, is much harder to bank into corners and unbank. Here, the Japanese bikes may have had to pay the price for their big powerful engines by being cumbersome in comparison to their European counterparts. It means that, for all their extra power, some Japanese superbikes have been slower over winding roads than their nimble, easy-handling European rivals.

In studying the suspension weaknesses, the team at Manchester pointed at the front telescopic fork as a prime offender. Apart from the wheel spindle, the two legs of the fork are joined only by a crossbar at the top, and each leg is effectively unsupported. The wheel is prevented from wobbling only by the strength of the individual legs—and they are often not very strong. Alternative front suspension systems have been devised, such as the hub-centre steering found on the Quasar where the wheel swivels on a pin in the centre of the wheel and the legs are fixed.

Nevertheless, although telescopic forks are far from perfect, the forks of today's superbikes have benefitted from considerable advances in spring and damper design. The development of damper units using an air component—'gas shocks'—have come close to realizing the ideal of a smoothly progressive damping rate, with gentle damping for small bumps hit at low speed to heavy damping for big, high speed bumps. Now an air-filled spring unit is replacing the present metal coil to give genuinely progressive damping.

Improvements in suspension quality have given the modern bike much better roadholding than its forebears. But even more significant than suspension improvements has been the steady progress in tyre technology. As much extensive research has been conducted on the tyres by the tyre manufacturers as any other part of the bike and the result is a range of tyres which give superb adhesion in both wet and dry combined with a reasonable wear rate, and they are improving every year.

Index

Numbers in italics refer to illustrations

Acoustic detectors, 46
Activity-time curves, 89–90
AD 500 airship, *154–6*
Addresses, computer, 19
Advanced Gas-Cooled Reactor (AGR), 40, *40*
Advanced Passenger Train (APT), 165, *166*
Advancing Blade Concept (ABC), 164
Aerotrain, French, 166
Air Service Movement Indicator (ASMI), 8
Airship Industries, 156
Airships, 153–6
  attractive fuels, 154
  non-rigid, 154, 156
  rigid, 154
  semi-rigid, 154
  vectored thrust, 154, *154*
Air Traffic Control (ATC), 6–8, *8*
  ASMI, 8
  computerization, 6–8
  DVOR, 8
  Instrument Landing System (ILS), 7, 8
  Microwave Landing System (MLS), 8
  radar, 6–8
  safe to land, 8
  stacking areas, *6*, 8
AK-47 rifle (Soviet), 112, *113*
Alarm sensors, 46–8
Alberti, Leo Battista, 31
Albion-Carmichael Firechief, 159
Albuquerque solar power station, *51*
ALCM (air-launched cruise missiles), 110, 111
Algorithms, 28, 32
All-Union Scientists' Research Institute of Current Sources, 50
American Cancer Society, 84
American National Bureau of Standards, 32, *33*
AMX-30 tank (French), 102, 104
Amplifiers, 73
Anaemia, 86
Analogue music encoding, 70–73, *72*
Animation, animated films, 52–4
  'Animal Farm', 52, *52–3*
  computers used in, 52, 54, *54*
Animation or rostrum camera, 52, *53*
Anti-aircraft missiles, 114
Anti-ballistic missiles (ABMs), 115
Anti-lock braking system, 178
Anti-tank weapons, 113
APDS projectile, 104
Apollo missions, 132, *132*, 134, *134*, *135*
Aqualung (or SCUBA), 55–6
  cylinders, 55
  demand valve, 55, *55*, 56, *56*
  harness, 55, 56
Arecibo radio telescope, 130
Ariane (space satellite launcher), 123–5, *123*, *124–5*
  future orders, 125
  improvements, 124–5
  three stages (L140, L33, H8), 124
  variants II and III, 124
Armalite AR-15 rifle (US), 112, *112–13*
Armalite M16 rifle (US), 108
Armoured personnel carrier (APC), 114
Armoured vehicles, 102–4
Armstrong, Neil, 135
Arthroscopes, 81
Artificial pacemakers (heart), 96–8
Artificial valves (heart), 95–6, *95–6*
Aspect ratio, increasing, 59
Aspirator, *99*
Athabasca oil mining (Canada), 42
Atlas-Centaur launcher, 124
Atom(s), 79, 130
  splitting the, 38
Atomic bomb, 38
Audio encryption units, *32*
Australian solar powered telephone link, 50
Autogyros, 162
Autorotation, 164

BBC *Ceefax* and *Orbit*, 25–6
Bean, Al, 134
Bell Aerospace GEV landing craft, *170*
Bell Telephone Laboratories, 129
Bernouilli principle, 171

Beta radiation, 90
Betamix C-7 videorecorder, *75*
Binary additions, 28
Binary notation, 19
Bipolar memories, 20
Bitumen, 41–2
Blood pressure, high, 86
Body scanners, 77–9
  CT scanners, 77–8, *77–8*
  NMR-imaging, 79
  ultrasound scanners, 78–9, *79*
Boeing Chinook HC Mk 1 helicopter, *162–3*
Boeing 'High Point' hydrofoil, 173
Boeing jetfoil, *172–3*
Boeing Tucumcari hydrofoil, 173
Boiling Water Reactor, 40
Bomb disposal units (EODs), 105–7
  'Marauder', 107
  urban guerrillas and, 105–6
  Wheelbarrow Mk 7 EOD Unit, *105–6*, 107
Bone scan, 88–9
Bourne, William, 117
Bowling alley, 57–8
Brain scanning, 77–8
British Hovercraft Corporation SR-N4, *168–9*, 170
Bridge-laying vehicles, 104
British Rail signal box, *66*
British Telecom, 27
Bullets/cartridges, 108–9, 112, *112*
  armour-piercing, *108*
  consumable-cased cartridge, 108
  effect geschoss, 109
  flechette cartridge, *108*
  hit probability, 108
  lead-cored rifle, *108*
  Löffelspitz, *109*
  pistol, 109, *109*
  spitzer, 109
  tracer, *109*

Calculators, electronic, 28–30
  alphanumerics, *28*, 30
  CPU, 28, 30
  exploded view of, *29*
  lcds (liquid crystal displays), *28*, 30, *30*
  led displays, 30
  seven segment display, *30*
  vf (vacuum fluorescent) displays, 30
Calder Hall (Nuclear) Reactor, 39
Cameras:
  animation or rostrum, 52, *53*
  colour video, *74*
  ENG, 22–4
  gamma, 88, 89, 90, *90*
  instant picture, 62–4
  matchbox-sized solid-state, 24
  space, 132–5, 146
Cancer treatment, 83, 84, 85, 90
Capacitation of spermatozoa, 100
Cape Canaveral, 123, 136, 138–9
Carburettors, CV, 180
Card access control unit, 48, *48*
Cars, vehicles:
  electric, 157–8
  hydrogen, 174–5
  in-car computers, 176–8
Cartridges *see* Bullets
Cassegrain focus, *126*, 128
Catheterization, *88*
Cathode ray tube, 11, 12, 17
Cavitation, 173
Cavity flow, 173
CCD (charge-coupled device) memories, 21, 24
Ceefax (BBC teletext), 25–6, 27
Central Processing Unit (CPU), 28, 230
Centrifuge, *145*
Chieftain tank (British), 102, 104
Chromatic aberration, 126
Chrominance, 12
Ciphers, ciphering, 31–3
  digital techniques, 32
  polyalphabetic, 31, *31*, 32
  pseudo-random sequence, 32
  running key, 31, *31*, 32
  substitution, 31
  transposition, 31
Citizens' Band Radio (CB), 9
  base station, 9
  hand portable ('walkie talkie'), 9
  mobile station, 9, *9*
CMOS (Complementary MOS), 20
Cold Lake huff and puff project (Esso's), *41*, 42
Collective pitch control, 162
Collimator, *90*
Colonoscopes, 80

Colour television, 11–12, 14, 15
Columbia orbiter, 138, 139, 140
  flight deck, *138–9*
  return flight path, *140–1*
Communications satellites, 13–15
  direct broadcasting satellites, 15
  ground stations, 15
  Intelsat, 14, 125
  launch sequence, *14*, 14–15
  OTS2: *13*, *14*
  Telstar 2: 15, *15*
Complementary addition, 28
Computer language, 19
Computer memories, 17, 19–21
Computerized tomographic (CT) scanners, 77–8, *77–8*, 79
Computers:
  animated films, 52–4, *54*
  in-car, 176–8
  printing, 16–18
  security system, 46–8
  teletext, 25, *26–7*
  Viking, 146
Concentration gradient, 86
Concorde, 171
Conrad, Pete, 134
Contrast angiography, 90
Control rods (nuclear reactor), 38
Coudé focus, *126*, 128
Cousteau, Jacques-Yves, 55
Crippen, Robert, 139
Cruise missiles, 110–11, 118
  ALCM (air-launched), 110, 111
  GLCM (ground-launched), 111
  Harpoon, 110, 111
  SLCM (sea-launched), 110, 111
  TERCOM system, 110–11
  Tomahawk, *110–11*, 111
Cryptanalysis, 31–2
Cryptography, 31
  high speed digital techniques, 32
  public key, 32
Cryptology, 31
Cryptosystems, 31–3
  computationally secure, 32
  DES, 32, *33*
  digital techniques, 32
Cyclic pitch control lever, 162

Daimler-Benz hydrogen car, *174*
Data Encryption Standard, 32, *33*
Deep-level 'atmospheric' railway, 165
Delivery vehicles, electric, 157
Demand valve (aqualung), 55, 56, *55–6*
Desmodromic system of valves, 180
DGV aircraft carrier, 121, 122
Dialysis *see* Kidney machine
Digital encoding of music, 73, *73*
Digitizer, *17*
'Dilemma' animation film, *54*
Direct broadcasting satellite, 15
Dish antennas, ground-receiving, 15, *15*
Discs:
  audio, 70–71
  video, 74, *75–6*
Distance Measuring Equipment (DME), 8
DNA (human interferon), 85
Dolby system of tape recording, 72
Drebbel, Cornelius van, 117
Dutch Loco 1306, HO model of, *65*
DVOR, 8
Dynamic scans, 89–90

EAROM (electrically alterable ROM), 20–21
EDS (electrodynamic suspension), 166–7
Edwards, M. L., 95
Edwards, Dr Robert, 99–100
Edwards Air Force Base, *140*
EETOI maglev train, German, 166
Effect geschoss bullets, 109
Effelsberg radio telescope, 130
Einstein, Albert, 36, 38
Electric vehicles/cars, 157–8
  batteries, *157*, 157–8
  hybrid petrol and, 158, *158*
  regenerative braking, 158
Electron beams (TV), 11, *11*, 12
Electronic calculators, 28–30
Electronic cars *see* in-car computers
Electronic digital cryptosystems, 32, *32*
Electronic easel, *54*
Electronic ignition, 176–7, 180–1, *181*
Electronic news gathering (ENG) 22–4
  one-man (live broadcast), 22–3, 23, *24*
  two-man, 22, *22–3*
  videorecorders, 23–4
Electrons, 38, 90, 131

EMS (electromagnetic suspension), 167
Enigma cipher machines, 31, 32
Enterprise orbiter, 138
EOD suit (bomb disposal), 106
EPROMs (Eraseable Programmable Read Only Memories), 20
Erythropoietin, 86
European Space Agency (ESA):
  Ariane, 123–5
  'Giotto' project, 125
  Spacelab, 140, 142–5
Explosive ordnance disposal (EOD), 105–7
  Wheelbarrow Mk 7 Unit, *105–6*, 107
Eye surgery, lasers used for, 37

F-111 swing-wing bomber, 111
Fallopian tubes, test-tube baby technique and, 99–101
Fast Breeder Reactor, *38–9*, 40
Fertilization, 99
  test-tube (*in vitro*), 99–101
Fibroblast cell interferon, 84
Field-effect transistor (fet), 20
Film news gathering, 22, *22–3*
Fire engines, 159–61
  foam tenders, 160
  pumps, 159
  Simon Snorkel fire-fighting unit, *159*, *160–1*
  'Snorkel' or hydraulic platform, 160
  super pumper complex, 160
  turntable ladder, 160
'First-pass' technique, 90
Fistula, 86
Fixed temperature detector, *48*
Flat plate collectors, 49
Fleet Ballistic Missile Submarine (SSBN), *118–19*
Fleuss, H. A., 55
Fluorescent collectors, 50
Flying bombs, radio-controlled, 110
FM (frequency modulation), 73
Focke-Achgelis Fa 61 helicopter, 162
Ford's experimental hybrid car, 158
Fort McMurray oil mining, 41–2, *42*
French high-speed trains:
  Aerotrain, 166
  TGV, *165*
Frequency dispersion techniques, *32*
Frog's egg, RNA injected into, *85*
Fuel injection systems, electronic, 176, 177, *178*

Gabor, Dennis, 34
Gagnan, Emil, 55
Galaxies, radio emissions from, 131
Galileo Galilei, 126, 128
Gamma camera, 88, 89, 90, *90*
Gamma radiation, 88
Gas chromatograph mass spectrometer, 148
Gastroscopes, 81, *81*, 82
Gemini Project (1965/66), 132
Genetic engineering, 83, 84–5
Geo-stationary orbit, 123
German high-speed trains:
  EE TO1: 166
  Transrapid 03 hovertrain, 166
'Giotto' (ESA project), 125
GLCM (ground-launched cruise missiles), 111
Glomerulus, 86
Gnomon (lunar tripod), *135*
Goodyear airships, *153*, 154, 156
Goonhilly dish antenna, *15*
GPMG (general-purpose machine gun), 113
Graf Zeppelin, 153, 154
Graphite, 38, 39
Gressor, Ion, 84
Grumman 698 aircraft, 121–2

Hagelin, Boris, *33*
Hagelin cipher machines, 32, *33*
Hale 5m reflector telescope, Mt Wilson, 128
Halley's Comet, 125
Hang gliding, 59–61
Harpoon cruise missiles, 110, 111
Harrier 'Jump Jet', *120–1*, 121
Hasselblad space camera, 132, *133*, *134*, 134–5
Hawker P1127 VTOL aircraft, 120, 121
Heart function/diseases, 95–8
  artificial pacemaker, 96–8
  artificial valves, 95–6, *95–6*
  catheterization, *88*, 90
  contrast angiography, 90

dialysis patients and, 86
heart block, 96, 97
normal operation of heart, 96, *97*
nuclear angiograms, 90
HEAT projectile, 104
Helical scanning, 74
Helicopters, 162–4
Advancing Blade Concept, 164
angle of attack, 162, 164
autorotation, 164
Boeing Chinook HC Mk 1: *162–3*
down draught, *164*
as microwave relay stations, 24
torque compensation system, 164
Helium, 153, 154
Heparin, 86
Hepatitis, 85, 87
Herschel Telescope, La Palma, *127*
Hesketh 1000cc superbike, *179*
Hi-fi systems, 70–3, *70–1*
amplifiers, 73
discs, 70–1
effect of micro-chip on, 73
loudspeakers, 71, 73
pick-ups, 71
radio-broadcasting, 72–3
tape-recording, 71–2
High frequencies:
audio (VHF), 72–3
video, 74–5
High-speed trains, 165–7
maglevs, 166–7, *167*
levitation, 165–6
power problems, 167
High Temperature Gas-Cooled Reactor (HTGR), 40
Hindenburg (zeppelin), 153, 154
Holography, 34–5
industrial uses, 35
laser light, *34*, 34–5, *35*
Horse Mesa Dam, Arizona, 50
Houston Space Center, USA
Mission Control, 145
Photographic Technology Division, 132, *134*
Hovercraft, 168–70
SR-N4, *168–9*, 170
Hovertrains, 166
Huff-and-puff technique, *41*, 42
Hybrid (petrol and electric) cars, 158, *158*
Hybrid solar/hydro-electric plant, 50
Hydrofoil, 170, 171–3
Boeing jetfoil, *172–3*
fully submerged, 171
types of hydrofoil system, *172–3*
vee-foil surface-piercing system, 171, 173
Hydrogen car, 174–5
fuel cell, 175
liquid hydrogen, 174
metal hydride, 175

IBA, 25
Oracle, 25, *25*, 26
Ignition:
electronic car, 176–7
superbike, 180–1, *181*
Image intensifiers, 128
In-car computers, 176–8
anti-lock braking system, 178
electronic ignition, 176–7
engine valves, 177–8
fuel injection systems, 176, 177, *178*
knock sensor, 177
providing the information, 176
solid state instrument panel, *177*
transmission, 178
Infantry weapons, 112–14
machine guns, 113
mortar attack, 114
portable missiles, 114
rifles, 112, *112–13*
rocket launchers, 113–14
sub-machine guns, 113
Infra-red light beams, 47
Infra-red pictures of weather, 151
Infra-red sensors, 47
Instant picture cameras, 62–4
Instrument Landing System (ILS), *7*, 8
glide slope aerial beam, *7*, 8
localizer aerial beam, *7*, 8
Intelsat, 14, 125
Interfacing (in-car computer), 176
Interference (laser light), 34, 35
Interference fringes, 130
Interferometers, 130, 131
Very Large Array, *129*, 130, *130*, *131*
Interferon, 83–5

genetic engineering, 84–5
synthetic production, 84–5
treating cancer with, 83, 84
virus infections controlled by, 83–4, *85*
In vitro fertilization, 99–101
Ionization detector, *48*
Ionosphere, 13, 14
Intelset (satellite network), 14
Isaacs, Alick, 83
Italian high-speed train, 165, 166

Jansky, Karl, 129
Japanese high-speed trains:
'bullet', 165
HSST, *167*
ML500: 166, 167
Japanese superbikes, *179*, 179–80, *181*
Jet (reaction) engines, 120
Jodrell Bank radio telescope, *129*, 130

Kalashnikov, Mikhail, 112
Karolinska Hospital, Sweden, 84
Kawasaki Z1300 superbikes, *179*
Kidney function:
nuclear investigation, 88
renogram, 89–90
Kidney machine (dialysis), 86–7
anaemia and high blood pressure, 86
cleaning the blood, 86
patient help, 86
portable, *87*
risk of infection, 86–7
side effects, 87
Kidney transplants, 86, 87
Kil plates, 86
Kingston valve, 117
Knee joint, endoscopic view of inside, *80*
Knock sensor, 177
Kodak Instant Print, 64
Kourou launching site (for Ariane), 123, *123*
Kruyer Tar Sands Development, 42

Lambda sensor, 177
Land, Edwin, 62
Laparoscope, laparoscopy, 99, 100
Large Scale Integration (LSI), 28
Lasers, 36–7
carbon dioxide, 37
communications, 37
endoscopic use of, 82
holograms produced by, *34*, 34–5, *35*
measuring with, 36–7
medical uses, 37, 82
military uses, 36, 37, 114
Philips Laservision, 75–6, *76*
phototypesetters and, 17
ruby rod, 36, *37*
LAW (British light anti-armour weapon), 113
Lenses, telescope, 126
Leonardo da Vinci, 117
Levitation high-speed trains, 166–7
maglevs, 166–7, *167*
Leopard 2 tank (German), 102, *102–3*
Lick Observatory refractor telescope, 126
Light(s):
coherent, 36, *37*
laser, 34–5, 36–7, *37*
surgical, 92, 94
wavelengths, 129
Light pen, *66*
Lindemann, Jean, 83
Linear synchronous motor, 166
Linoscreen VDU terminal, 18, *18*
Liquid hydrogen, 154, 174
Lithium iodine battery, 97, *97*
Löffelspitz bullets, 109, *109*
Looping Racer, 68
Loudspeakers, 71, 73
Lovell, Sir Bernard, 130
Lowell, Percival, 146
LTA craft *see* Airships
Luminance, 12
Luminescent greenhouse collector, 50
Lung test, 89

M60 tank (US), 102
M72 rocket launcher (US), 113
M551 Sheridan tank (US), 104
McDonnell Douglas Astronautics Co., 110
Machine guns, 113
Maglevs (magnetic levitation high-speed trains), 166–7
EDS, 166–7
EMS, 167, *167*

Magnet motors (model train), 65
Magnetic bubble memories, 21
Magnetic reed detector, 46
Magnetic scanning, 79
Magnox nuclear reactors, 39, 40
Maiman, Theodore, 36
Manned Manoeuvring Unit (MMU), *143*
Mariner space probes, 146
Mars, Viking Mission to, 146–9
Marshall Space Flight Center, 138
Maurer space 'movie' camera, 132, *133*
Mercury Project (1962), 132
Metal hydride, 175
Meteostat 1 weather satellite, *152*
Microelectronics, lasers used in, 37
Microprocessors, in-car, 176–8
Microswitch, 46
Microwave detectors, 46–7
Microwave Landing System (MLS), 8
Microwelding, 37
MILAN (Missile d'Infanterie Leger Anti-Char) anti-tank missile, 113
Milky Way, 129, 131
Minuteman III missiles, 115, *115*
MIRVs (multiple independently-targeted re-entry vehicles), 115–16
bus route, 115
limited areas, 115
terminal guidance system, 116
Missiles:
ABMs, 115
anti-tank, 113–14
cruise, 110–11, 118
Minuteman III, 115, *115*
MIRVs, 115–16
MX, 116
Polaris, 118, *118–19*
portable anti-aircraft, 114
Trident, 118
ML 81mm L16 mortars, 114
Mobile two-way radio telephone, 9, *9*
Model railways, 65–6
command control, 65–6
magnet motors, 65
power transistors, 65
Moderators, 38, 39
Molecules, 130
Mont Louis solar furnace, France, *51*
Moon, photos of landings on, 135, *135*
Moorfields Eye Hospital, 84
Mortars, 113, 114, *114*
MOS (Metal Oxide Silicon or Metal Oxide semi-conductor) transistors, 20
Mt Palomar reflector telescope, 128
MRVs (multiple re-entry vehicles), 115
MTC control system, 66
Multiplexing, 47, *47*
Music cassettes, 71–3
MX missile system, 116
Myocardium, 96

NASA (National Aeronautics and Space Agency):
camera regulations, 132
'getaway specials', 140
Nimbus 5 weather satellite, *151*
Nimbus B Tracking System, *151*
Spacelab built for, 142–5
Space Shuttle, 136–41
Synchronous Meteorological Satellite, *150*
NATO, 108, 113
Natural gas, 154
Nautilus, USS (nuclear submarine), 118
NBD (norbonadiene), 49
Nephrons, 89
Neutral buoyancy, 117
Neutron stars, 131
Neutrons, 38, 40, 79
Nimbus B Tracking System, *151*
Nova/supernova explosions, 131
NTSC (National Television Systems Committee), 12
Nuclear angiograms, 90
Nuclear fission, 38
Nuclear magnetic resonance imaging (NMR), 79
Nuclear medicine, 88–90
beta radiation, 90
bone scan, 88–9
dynamic scans, 89–90
lung test, 89
nuclear angiograms, 90
static scans, 89
Nuclear reactors, 38–40
Advanced Gas-Cooled, 40, *40*
Boiling Water Reactor, 40
controlling chain reaction, 38

cooling reactor core, 39–40
Fast Breeder Reactor, *38–9*, 40
High Temperature Gas-Cooled Reactor, 40
Magnox Reactor, 39, 40
Pressurized Water Reactor, 40
radiation from, 38–9
Nuclear submarines, *117*, 117–18, *118–19*
Nuclear warheads, 111, 118
MIRV, *115*, 115–16
see also Missiles

OB (outside broadcast) units, 22, *24*
Objectives, telescope (lens and mirror), 126
Occidental Petroleum, 42
Odeillo solar power station, *49*, 50
Oil mines, 41–2
Fort McMurray, 41–2, *42*
huff-and-puff technique, *41*, 42
Oleophilitic material process, 42
Operating theatre, 91–4
clean room technology, 92
fast turnover, 91
operating tables, 92, *94*
preparing for surgery, 91
pure air supply, 92, *93*
surgical light, 92, 94
total body exhaust system, 92
transfer system, 91
Optical telescopes, 126–8
focus (prime, cassegrain and coudé), *126*, 128
reflector (mirrors), 126, *127*, 128, *128*
refractor (lenses), 126
Schmidt, 128
Space Telescope, 128
Oracle (ITV teletext), 25, *25*, 26, 27
Orbit (BBC teletext), 25–6, 27
Orbital industries, 144
Organ transplants, interferon used in treatment of, 85
Orient Express rollercoaster, Kansas City, 67, 68
OTS 2 satellite, *13*, *14*
Outdoor movement detector, *48*

Pacemakers, artificial heart, 96–8
external, 96
implanted, 96–7
lead(s), 97, *98*
phone check, 98
pulse generator, 96–7, *97*, *98*
stand-by or demand type, 97–8
PAINT computer system, 54
PAL (Phase Alternate Line), 12
Parallel addressing, 19
Pegasus engines, 121
Petri dish, 99
Philips Laservision, 75–6, *76*
Photo-electric conversion of solar energy, 49
Photo-electric detectors, 46
Photography:
holograms, 34–5
infra-red weather pictures, 151
instant picture, 62–4
space, 132–5
Photons, 36, *37*, 49
Phototypesetting, computerized, *16–17*, 17–18
Pickups, *70*, 71
Pilot drag, 59
Pinspotter, automatic, 57–8
Pistols, 113
Plaintext, 31, 32
Polaroid instant-picture cameras, 62
special film backs, 64
SX-70: *62*, *63*, 63–4
ultrasonic echo-sounding system, *63*
Pollution, 174
Polyp in colon, removal with endoscope, *80*, 81
Polysilicon, 20
Power transistors (model train), 65
Pressurized Water Reactor (PWR), 40
Prestel (viewdata system), *25*, 27
Printing, computers in, 16–18
mechanical machines, 16–17
phototypesetters, 17–18
Printouts, *17*
PROMs (Programmable ROMs), 20
Protons, 38, 79
PT-76 tank (Russian), 104
Pulmonary embolism. 89
Pulsars, 131
Pulse generator *see* Pacemaker

Quasars, 130, 131
Quasistellar objects, 131
Queen Elizabeth II (liner), 171

R150 airship, 156
Radar:
  ATC (primary and secondary), 6–8
  cruise missiles, 110, 111
  submarine, 118
Radiation, 38–9
  beta, 90
  gamma, 88, 89, *90*
Radio:
  beacons, 6, *6*
  beams, *7*
  broadcasts, 13, 14, 72–3
  Citizens' Band, 9, *9*
  frequency modulation (FM), 73
  VHF, 72–3
  VLF communications with
    submarines, 118
  waves/wavelengths, 129, 130–31
Radioactive nuclear waste, disposal of,
  38, 39
Radio galaxies, 131
Radionuclides, 88, *88*, 90
Radiopharmaceuticals, 88, 89, 90
Radiosondes (weather balloons), 152
Radio stars, 131
'Radio static', 129
Radio telescopes, 129–31
  different wavelengths, 130–1
  dish development, 129–30
  radio sources, 131, *131*
  Very Large Array interferometers,
    129, 130, *130*, 131
Radio tuners, 70, 72–3
RAF Lakenheath, 111
RAF Upper Heyford, 111
RAM (Random Access Memory), 19–21
  dynamic, 19, *19*, 20
  static, 19–20
RBS 70 (Bofors) anti-aircraft missile, 14
Reber, Grote, radio telescope of, 129–30
RCA Selectavision, 76
Recording of music, high fidelity, 70–3
Recovery vehicles, 104
Regenerative braking, 158
Regenerative steering, 104
Renogram, 89–90
The Revolution roller coaster, Blackpool,
  68
Rifles, 112, 113
  AK-47 Kalashnikov, 112, *113*
  Armalite AR-15, 112, *112–13*
  Armalite M16, 108
Robots (industrial), 43–5, *44–5*
  continuous path operation, 44
  control system, 43, 44–5
  feedback element, 45
  hydraulic power unit, 43
  mechanical unit, 43
  point to point operation, 44
  servo valve, 43, 44, 45
Rocket-launchers (military), 113–14
Rockets (space):
  Ariane, 123–5
  satellite three-stage, 14–15
  Space Shuttle, 136–41
  Titan III Centaur, *146–7*
Roller coasters, 67–9
Rolls Royce jet engines, 120, 121
  RB108, 121
ROMs (Read Only Memories), 20, 30
Rotor cipher machines, 31–2
Rogallo Delta-shape hang gliders, 59
Rotor disc (helicopter), 162, 164
RTV hovertrain, 166
Ryle, Sir Martin, 130

SA-7 anti-aircraft missile (Soviet), 114
Safe limpet, 118
St Jude Medical bi-leaflet valve, 95–6,
  95–6
Salt River project, Arizona, 50
Sandia solar power tower, New Mexico,
  150
Satellites, 118, 123, 128, 131
  Ariane, 123–5
  communications, 13–15, 125
  direct broadcasting, 15
  geostationary, 15, 123, 150–1
  Intelsat, 14, 125
  polar-orbiting, 152
  Space Shuttle, 15, 140
  weather, 150–2
  see also Space satellite launchers
Saturn's rings, 126

Scanning, scans, 11–12, 17
  double-interlaced, 12
  dynamic, 89–90
  helical, 74
  radio nuclide, 88
  static, 88–9
  see also Body scanners
Schmidt telescopes, 128
Scorpion tank (British), 104
'Scrubbing' air, 118
SCUBA, 55–6
Sea Harrier V/STOL aircraft, *120–1*, 122
SECAM (Système en Couleurs à
  Mémoire), 12
Security systems, integrated, 46–8, *46–7*
Seisometer, 146
Serial addressing, 19
Servo valve, 43, 44
Shale oil, 41–2, *42*
Ship Inertial Navigation System (SINS),
  118
Short SC-1 delta wing research aircraft,
  121
Shunt, 86
The Shuttle Loop, *69*
Siding Spring 3.9m telescope, *128*
Sikorsky Company, 162, 164
Silicon chips (microchips), 19, 20, 73
  calculators, 28–30
  CCD, 24
  in-car microprocessors, 176–8
  LSI, 28
  number of AEGs, 28
  satellites, 14
Simon Snorkel fire-fighting unit, *159*,
  160–1
Sino-atrial (S-A) node, 96, *97*
Sirius, white dwarf star accompanying,
  126
Skylab, 132, 135
SLCM (sea-launched cruise missiles),
  110, 111
'Sniffer' (explosive detector), 106
Sodium solar tower, *50*
Solar flare, 131
Solar power, 49–51
  central receiver collector, 49–50
  collector system, 49
  fluorescent collectors, 50
  parabolic dish concept, 49, *49*
  photo-electric conversion, 49
  sodium, *50*
Solid state instrument panel, *177*
Sony videorecorders, 23–4, *75*
Soviet hydrofoil, *171*
Space cameras, 132–5
  exposure charts, 134
  Hasselblad, 132, *133*, 134
  limited regulations of NASA, 132
  Maurer 'movie', 132, *133*
  Nikon F3, *133*
  photos of lunar surface, 135, *135*
  space guidelines, 134
  special film and processing, 132
  spot meters, 134
  training astronauts to use, 132, 134
  viewfinding, 134–5
Spacecraft, space launchers:
  Ariane, 123–5
  Atlas-Centaur, 124
  Space Shuttle, 123, 124, 136–41
  US Delta, 124
  Viking, 146–9
Spacelab (space station), 140, 142–5
  laboratory module, 142, *142–3*, 144
  MMU space suits, *143*
  outside equipment, 144
  pallets, 142, *142–3*, 144
  payload specialists, 144, 145
  power sources for, 145
Space probes, 140
Space Shuttle, 15, 123, 124, 134, 136–41
  Columbia's return flight path, *140–41*
  heat shield tiles, 138
  launch, 136, *138*, 138–9
  multiple use, 136
  Nikon F3 camera to be used in, *133*
  orbiters, 136, 138, *138–9*, 139, 140
  regular operation, 140
  satellites, 140
  Spacelab in cargo bay of, 142–5
  stages of typical mission, *136*
  test program, 136, 138
Space sickness, 144
Space Telescope, 128, 144
Spanish high-speed train, 165
Spectroscopes, 128
SPG-9 rocket launcher (Soviet), 113

Spiral galaxies, 131
Spot-welding guns, 43
S-tank, Swedish, 104, *104*
Starr-Edwards ball valve, 95, *95*
Stars, radio emissions from, 129, 131
Static scan, 88–9
Steptoe, Patrick, 99, 100, 101
Stereo, 70–3
Stereoscope, 34
Stoner, Eugene, 112
Strip-mining (oil) techniques, 42
Strontium-90, 38
Sturmegewehr (automatic rifle), 108
Sub-machine guns, 113
Submarines, 117–19
  armoury capacity, 118
  diesel-powered, 117–18
  nuclear-powered, *117*, 118
  SINS, 118
  SSBN, *118–19*
  Trident missile, *119*
Sugar residues, 85
Sun, radio emission from, 131
Sunspots, 131
Superb, HMS (nuclear submarine), *117*
Superbikes, 179–81
  CV carburettors, 180
  ignition, 180–1, *181*
  power boost, 179–80
  suspension and stability, 181
  tyre technology, 181
  valve gears, 180, *180*
Superconductivity, 167
Superovulation, 100
Surgery *see* Operating theatre
Surgical light, 92, 94
Swedish robot, *43*
Syncrude (oil), 42
System 4 (computer animation), *54*

T-72 MBT tank (Soviet), 102, 113
TALCM (Tomahawk ALCM), 111
Tanks, 102–4
  anti-aircraft, 104
  German Leopard 2, *102–3*
  guns, 104
  main battle, 102
  reconnaissance, 102, 104
  regenerative steering, 104
  tracks, 104
Tape recorders:
  cassettes, 71–2
  Dolby system, 72
  open-reel, 71
  multi-track, 70
  see also Video
Tar sands, 41–2
Technetium, *88*, 89
Teledyne turbojet engine, 110
Telephone cables/lines, 14
Telescope:
  optical, 126–8
  radio, 129–31
  Space, 128, 144
Telesoftware, 26
Teletext, 16, 25–6, *26–7*, 27
  Ceefax, 25–6
  Oracle, 25, 25, 26
  Orbit, 25–6
  telesoftware, 26
Television broadcasts, 13, 14, 15
Television information services, 25–7
  teletext, 25–6
  viewdata, 26–7
  see also Colour television
Telstar 2, 15, *15*
Temperature detectors, 46
Temperature rate-of-rise detector, *48*
TERCOM (terrain contour-matching)
  system, 110–11
Terminal guidance system, 116
Terrain matching system, 116
Test tube baby, 99–101, *100–101*
Thermal soaring, *59*
Tilt nacelle concept, 122
Time base correct (videotape), 24
Titan III Centaur rocket, *146–7*
Tomahawk cruise missiles:
  AGM-109 (air-launched: TALCM), 111
  BGM-109 (sea-launched), *110–11*, 111
  BGM-109B (ground-launched), 111
Torpedoes, 118
'Total body exhaust' system, 92
TOW (tube-launched, optically tracked
  wire-commanded) missile, 114, *114*
Transceivers:
  base station, 9
  hand portable, 9

mobile stations, 9, *9*
Transistor flip-flops, 19, 20
Transistorized controls (model train), 65
Transrapid O3 hovertrain, 166
Tree form data base, 27
Triangulation (superbike), 181
Tri-cart (hang gliding), 60
Trident missiles, 118
Trident missile submarines, *119*
Trip computers, *177*
Tubules (kidney), 86
TV cameras, 12
  ENG, 22–4
Tyre technology, 181

UDMH, 124
Ultrasonic detectors, 46, 47
Ultrasonic echo-sounding system
  (Polaroid), *63*
Ultrasound body scanners, 78–9, *79*
  'real-time', 79
Uranium, 38, 39–40
  -235, 58
  -238, 38
  enriched, 38, 40
Uranus, *127*
Urban warfare, 109
  EODs and, 105–7

Valve gears, superbike, 180, *180*
Ventilation, 173
VHD video system, 76
VHF radio, 72, 159
Video camera, colour, *74*
Viewdata, 16, 25, 26–7
  Prestel, *25*, 27
  tree form data base, 27
Video (recording and replay) systems,
  23–4, 74–6
  Betamix C-7 video recorder, 75
  disc-based, 75, 75–6
  frequency problems, 74–6
  helical scanning, 74
  Philips Laservision, 75–6, *76*
  RCA Selectavision, 76
  tape-based systems, 23–4, 74–5
  time base correct, 24
  time code, 24
  U-Matic system, 23, 24
  VHD, 76
Viking Mission (to Mars), 146–9
  gas-exchange experiment, 147–8
  labelled-release experiment, 148
  lander, 146, 147, *148–9*
  orbiter, 146, *148*
  pyrolitic release experiment, 148
  Titan III Centaur rocket, *146–7*
Viral infections, interferon's prevention
  of, 83–4, *85*
Visual display units (VDUs), 47, 48
VLF radio communications, 118
Volvo's assembly shop robot, *43*
VS 316A (later R-4) helicopter, 162
V/STOL (vertical or short take-off), 122
VTOL (vertical take-off) planes, 120–2
  Hawker P1127, 121
  nautical aspect, 121–2
  Sea Harrier, 122
  vectored thrust, 154
  Yak Forger (Russian), 122

Walkie-talkies, 9, 159
Weather satellites, 150–2
  geostationary, *150*, 150–1, *152*
  infra-red pictures taken by, 151
  polar-orbiting, 151
  radiation measurement, 152
  visual pictures taken by, 151
  wind velocity determination, 151–2
Wheelbarrow Mk 7 EOD Unit, *105–6*, 107
White blood cell interferon, 84

X-ray bomb disposal equipment, 106
X-ray body scanners, CT, 77–8, *77–8*, 79
X-rays, conventional, 77, 88

Yak Forger VTOL aircraft, 122
Yerkes Observatory refractor telescope,
  126
Young, John, 134, *135*, 139

Zero 1 control system (model train), *65*,
  66, *66*
Zero gravity, 144, 145
Zero or null flux guidance, 166
Zircaloy, 40
ZPG3W naval airship, 156

184

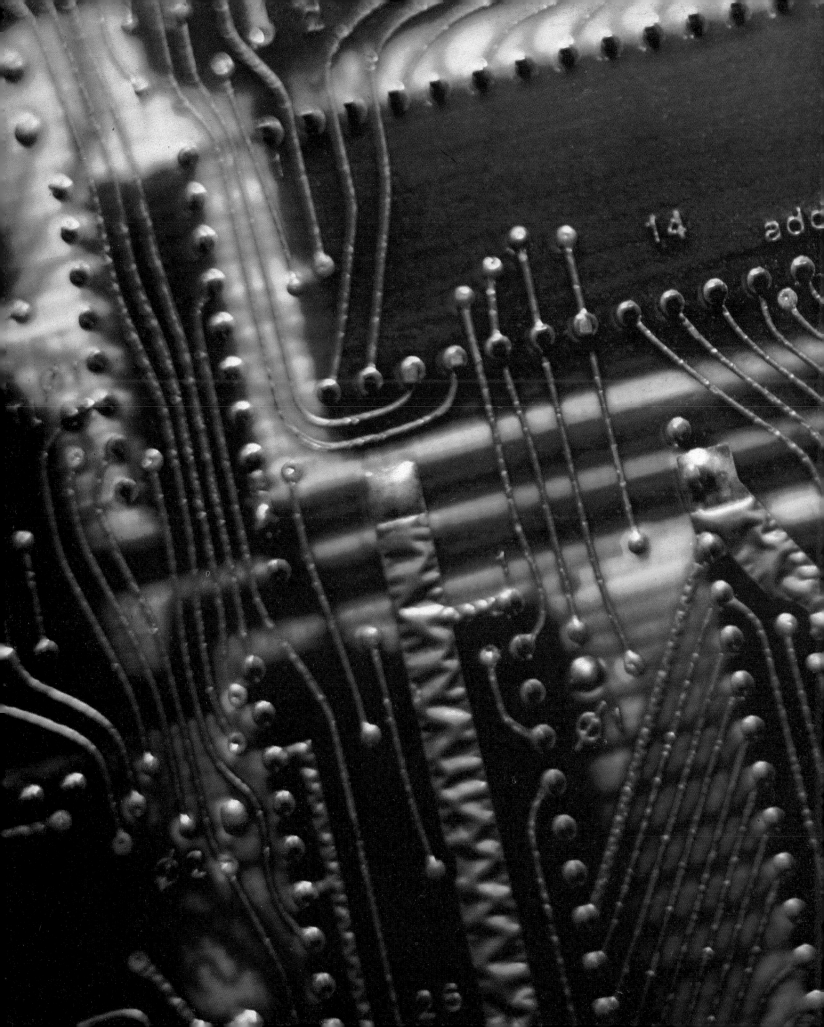